WIND ENERGY
AND THE ENVIRONMENT

EDITED BY
D T SWIFT-HOOK

Peter Peregrinus Ltd. on behalf of the Institution of Electrical Engineers

Published by: Peter Peregrinus Ltd., London, United Kingdom

British Library Cataloguing in Publication Data

Wind energy and the environment.
1. Environment. Effects of wind power
I. Swift-Hook, D.T.
333.7′1

ISBN 0 86341 176 2

Printed in England by Short Run Press Ltd., Exeter

IEE Energy Series 4
Series Editors: M. Barak and
Professor D. T. Swift-Hook

WIND ENERGY

AND
THE ENVIRONMENT

Other volumes in this series

Volume 1	Electrochemical power sources M. Barak
Volume 2	Renewable energies: sources, conversion and application P. D. Dunn
Volume 3	Combined heat and power generating systems J. Marecki

Contents

About the European Wind Energy Association

The EWEA was formed as an international Association to unite Europeans and others who are active in the field of wind energy and to promote exchanges in wind energy research, development and expansion, not just for the benefit of specialists but also for the benefit of the general population. An inaugural meeting took place in 1982 when Professor Swift-Hook, who was Chairman of the already established British Wind Energy Association, called together representatives of other National Associations that were being formed throughout Europe and the EWEA has grown steadily since then.

The Association advises National Governments, the European Communities and other bodies on the planning and administration of wind energy activities and it establishes authoritative opinions on technical and other matters through meetings, conferences and publications.

Although it is a professional European body (duly incorporated in Brussels and requiring a degree or equivalent for Ordinary Membership), the Association is essentially outward looking and encourages Associate and Honorary Membership from individuals and organisations who are concerned with wind energy at any level and in any part of the world. There are close links with the American and Chinese Wind Energy Associations.

About the IEE Energy Series of Books

The Institution of Electrical Engineers is the largest professional body based in Europe with a world-wide membership of more than 100,000 men and women in the fields of electrical power engineering, communications, electronics, computing, software engineering and control.

The IEE is a leading technical publishing house which produces a wide range of journals, books and other publications covering management, design, education and many fields of science and technology within and beyond the boundaries of the profession. Perhaps its most well known publication is the IEE Wiring Regulations which are used world-wide but its many Series of Books cover such diverse fields as Management, Measurement, History, Telecommunications, Computing, Power Engineering, Broadcasting, Materials, Radar, Control and Craft Studies.

The Energy Series of books is concerned with all methods of electrical power generation but with particular emphasis on the renewables. The Series is edited by Dr L V Divone who is the Director responsible for solar electric technologies (including wind power) in the US Department of Energy and by Professor D T Swift-Hook who recently retired from the UK Central Electricity Generating Board and is Director of the Centre for Applied Research at King's College, London as well as being closely concerned with wind energy in Britain and Europe.

Foreword

Environmental and institutional concerns are the only ones now standing in the way of the wide-spread exploitation of wind energy. In this volume, the European Wind Energy Association presents a series of contributions which will help to resolve many of these outstanding problems.

Wind energy has long been recognised as benign becauses it is safe, non-polluting and does not deplete the world's energy resources. Now it can also claim to be cheaper in many parts of Europe than any other method of generating electricity that can be installed. On good windy sites, where plenty of wind can be expected, wind turbines are now competitive with fossil-fuelled or nuclear plant for power generation. Detailed costings are set out in the first paper in this volume which also describes some of the technology on which these claims are based and the performances achieved by such plant which has been exported successfully from Europe to California .

Any electrical power utility will naturally want to install the type of power plant that will generate the cheapest electricity. Wind farms now provide the most economical method of generation and so active consideration is being given to their wide-spread use. Questions are naturally being asked about technical and financial matters such as reliability and the costs of operation and maintenance and answers to many of those questions are emerging from the wind farms already in service. But by far the strongest reactions from the utilities and the biggest question-marks they are now raising are to do with environmental problems and siting considerations.

Wind farms cover considerable areas of countryside and although they occupy only a tiny fraction of the land (only a per cent or two), their presence will be noticeable over the whole area. If they make a noise, they may be heard and if they interfere with radio or television reception, there will be objections. They will certainly be visible at quite a distance and the public will need reassurance about many other matters : no one wants to be hit by a flying wind turbine blade.

Cautious responses and expressions of concern do not necessarily indicate any lack of enthusiasm for wind energy on the part of the utilities. They appreciate only too well the advantages offerred by a method of generation that produces no acid rain or carbon dioxide and no radioactivity, a renewable source of power that does not deplete world reserves. It is simply that they have long experience with environmental problems and public objections caused by their present types of power plant.

Wind farms could not blight the forests of Germany nor the lakes of Norway and they could not conceivably produce another Chernobyl but that does not mean that they will have no environmental effects of their own. The wind energy community is tackling these environmental problems - and tackling them with the same enthusiasm that developed the present machines so successfully. Wind turbine technology has many facets, from aerodynamics and mechanics to control and electronics. Civil engineering, mechanical engineering, electrical engineering and aeronautical engineering - many fields of technology are involved with some mathematics , physics and materials science thrown in. The environmental problems of wind energy cover even wider fields, not only in science and technology but on less quantifiable subjects such as public acceptability and legal constraints.

Noise restrict the siting of wind farms and several papers in this volume deal with the problems of noise : noise measurement, noise reduction and noise propagation. It will be just as challenging to develop low-noise rotors as it has been to produce reliable and economic ones. The aero-space industry has been successful in that direction with turbo-fans for aircraft and there are already signs that the wind industry will be able to follow suit. New methods of noise measurement need to be developed and revised recommended practices are included in the volume along with measurements made on specific machines.

Electromagnetic interference is probably at its most public when it affects TV, but radio and RADAR also need to be considered. Papers covering all these fields are included. Safety is vital with its legal as well as emotional implications : although wind turbines are likely to be built in somewhat remote and inhospitable parts of the countryside, the prospects and consequences of major structural failures need careful consideration. A paper on safety design codes to avoid structural failures is complemented by one on the blade throws that could result if they occur.

Public acceptability and legal constraints will ultimately determine the future of wind energy. In addition to the measurable parameters (such as audible noise level or microwave scattering cross-section) there are less tangible features to be explored. Legal constraints are often not so clear cut, especially in new situations when they may conflict with common sense. It is important to build up a body of experience and case histories against which to judge new developments. Several papers deal with public attitudes in specific European locations and there is a detailed review of legal constraints on wind turbines within the CEC.

This volume covers almost every important aspect of the environmental and legal issues facing wind energy today. The contributions presented here originated from a European Wind Energy Association Workshop convened in London by the Editor in March 1988. The mood of that Workshop was generally buoyant : it was agreed that ways forward could be seen in every area and that the problems do not appear to be insuperable. Although much remains to be done, wind energy can move forward confidently as the cheapest and safest method of generation, which is also non-polluting and renewable, and whose environmental problems and legal constraints do not seem to be insurmountable.

D T Swift-Hook,
Editor

Chapter 1

The technical and economic status of wind energy

D. Lindley and D. T. Swift-Hook

SYNOPSIS Wind energy has long been recognised as safe and non-polluting but now it can also claim to be the cheapest method of generating electricity that can be installed. Generation costs depend not only upon the initial capital cost per average kilowatt of the wind turbine and associated plant but also upon operating and maintenance costs. The average power output depends directly upon the % availabilities of the wind and of the wind turbine, as well as on machine performance. The cost of energy then depends upon the interest charged on the capital and the amortisation of the capital over the life of the plant. The paper describes how the costs of energy depend upon each of these parameters. Recent sales of European wind turbines into California have shown that prices in quantity production have fallen dramatically and that good performances can be achieved with high reliability for low operating and maintenance costs. Actual cost and performance figures are quoted for the first year of operation of one wind farm in California comprising twenty British machines. "Turn-key" capital costs are now less than £700 per rated kilowatt, average machine availability exceeded 95% and an annual load factor of 28% was achieved on a site with an annual-average wind speed of 7.8 m/s at hub height. On windy European sites, these figures give generation costs 20% less than nuclear or coal-fired generation. Despite the fact that the wind does not blow all the time, wind turbines can provide firm capacity on a power system. There therefore seem to be no technical or economic barriers to the wide-spread installation of wind power plant. The only remaining problems are environmental and institutional.

1. INTRODUCTION

Over the last fifteen years, substantial programmes of research and development in wind energy have been undertaken in many countries. Undoubtedly the biggest investment has been in California where, quite apart from research and development funding, tax credits amounting to nearly $1,000,000,000 have been awarded to wind farm constructors. Wind turbine manufacturers from Belgium, from Britain, from Denmark, from the Netherlands have all competed successfully with local American suppliers.

Undoubtedly, many of the early machines which were financed with American subsidies were unreliable and costly to maintain. A number of the less successful companies have gone out of business. Three or four years ago wind turbine costs were nowhere near economic and it was only tax subsidies which enabled them to be sold. Nevertheless, these tax incentives were dramatically successful in encouraging quantity production and, not surprisingly, prices have fallen. It can now be claimed (1) that , on good windy sites, wind turbines represent the cheapest method of generating electricity that can be installed.

2. PRACTICAL EXPERIENCE WITH A MODERN WIND FARM

To substantiate such a claim realistically, one must first be able to point to practical, working systems available commercially for which detailed operational results can be quoted and for which actual costs are known. One must then use appropriate methods of costing to calculate the cost of energy.

One wind farm for which details are available (2) came into operation on 1st January 1986. Results for the first complete year of wind farm operation are now available and form a useful basis for judging the performance that can now be acheived with modern European wind turbines.

The twenty machines ordered in April 1986 for this wind farm were supplied by the Wind Energy Group Ltd (jointly owned by British Aerospace plc and Taylor Woodrow Construction Ltd). They were manufactured and tested in the United Kingdom while US Windpower, the world's largest wind farmers, completed the site works in California, about 60 km east of San Fransisco. The MS-2 design (2)(3)(4) has an electrical output of 250 kW at the rated wind speed of 12.5 m/s. The 25 m diameter rotor has 3 blades each with full pitch control mounted up-wind of the structurally soft steel tube tower with active yaw .

In the first year of operation from 1st January to 31st December 1987 the wind generated 11.5 GWh, corresponding to an annual load factor of 28 %. The average wind speed was around 7.8 m/s at hub height, slightly below the average wind speed over the previous 8 years of 8 m/s.

Wind farm availiability measured in terms of hours out of service was 95 % in line with availiabilities reported for Californian wind farms in previous years (5). Wherever possible maintenance is carried out during non-windy periods so effective operational availiability and system peak availiability (which determines thw firmness of the capacity) are somewhat higher than 95 %

3. ECONOMICS IN EUROPE

The WEG wind farm in California was able to take advantage of tax credits there, but the experience of building, financing, operating and maintaining it together with the information available from the Californian Energy Commission on the operation of all other Californian wind farms indicate that wind farms are financeable there on an entirely commercial basis at sites with reasonable winds above 7.5 m/s (2).

Many parts of Europe with high land, especially in the North and the West have very good wind regimes. Studies for the Commission of European Communities have shown that very many sites are available, as many as 400,000 even allowing for all likely siting restrictions.

It will be much cheaper for European manufaturers to build wind farms in their own countries. WEG, for example, are now prepared to quote prices as low as £600/kW on a turn-key basis for wind farms on suitable sites in the UK and other European manufacturers claim to be competitive.

4. OPERATION, MAINTENANCE AND PLANT LIFE

Both operation and maintenance should be cheaper in many European countries where labour rates and equipment hire are both substantially less than in California. In the first year, operation and maintenance costs have averaged about 0.6 c/kWh for the WEG wind farm in California, and that should correspond to 0.2 p/kWh in Europe. The majority of faults in this initial period were of the "infant mortality" type that are unlikely to re-appear during the life of the plant. Electronic software and hardware problems accounted for 51 % of all faults, while electrical wiring, connection and components where responsible for a further 35 %. Only 5 % of all faults had a mechanical cause. Operation and maintenance costs can therefore be expected to fall significantly below the 0.2 pkWh figure throughout the life of the plant.

Plant life must, of necessity, represent a projection. Most engineering design codes specify at least 25 years life but early failures in the first year or two have been experienced by a number of manufacturers. The experience gained from such failures gives added confidence in more recent designs. For example US WEG Inc have leased back for an initial 12 years the wind farm which they sold to a major US investor . During that time they expect to recover all their costs and to make a profit on the sale of electricity to the Californian utility, Pacific Gas & Electric Company, under a 25 year contract. It therefore appears that a 25 year life can be projected with some confidence but it would be wise to make provision for some major refurbishments, equivalent say to re-blading machines half way through their life.

These provisions for refurbishments will balance to a greater or lesser extent the reductions to be expected in annual O&M costs. The changes in either direction are difficult to project accurately but are both likely to represent less than 10 % of the total cost of energy. A reasonable initial assumption will therefore be that the amounts cancel and that if the initial annual O&M charge of 0.2 p/kWh indicated by WEG's first year of operation is maintained, it will be sufficient to provide for major replanting, such as replacing every wind turbine blade at some time during the life of the plant.

5. RATES OF RETURN

In order to use these various figures to quote a cost of energy, a rate of return on capital must be specified together with a figure for inflation. The simplest approach is to express all costs in present day terms and to uses a rate of return expressed in real terms ignoring inflation (roughly the money interest rate less the inflation rate).

Interest rates in many European countries are notoriously volatile, but most OECD Governments require a real rate of return on capital of around 5% for major long-term investments such as power plant. 5% is the figure used by the International Energy Agency (6). As this same figure happens to apply to commercial funds available in the UK today, it is a convenient one for us to take.

Over 25 years life, the rate of amortization (4%) and the real interest rate (5%) are comparable. The combined charge rate is then levelised at 7.1% (1).

6. COST CALCULATIONS

The cost calculations required have been set out in some detail by the British Wind Energy Association (1) and one or two refinements can be noted in passing. Decomissioning cost and scrap value will more or less cancel each other out (7). Interest during construction is normally charged on major projects but experience with wind farms is that they are earning revenue within a few months of the start of construction. The WEG wind farm, for example, was ordered in April 1986 and completed by the end of December 1986. With such short construction times, any interest during construction is likely to be borne by the constructor and included in the cost of the wind turbine itself.

It can be noted that a measurement of the total kWh generated in a year and hence of the over-all load factor (28%) embraces machine availability and other losses as well as the nominal wind load factor. Improvements in any of these various factors will improve the over-all load factor. For example, an 8.5 m/s annual average wind speed should give 20% more annual energy from the wind turbine than the 7.8 m/s experienced by the WEG wind farm in 1987 and the over-all load factor would be increased from 28% to nearly 34%.

With a load factor of 34%, the quoted price of £600 per rated kilowatt corresponds to £1,750 per average kilowatt or just over 20 p/kWh/a. At an annual charge rate of 7.1% (to amortize over 25 years life with a real rate of return of 5%), this gives a cost of 1.43 p/kWh to which must be added the O&M costs of 0.2 p/kWh, giving a total generation cost of less than 1.7 p/kWh.

It is a complicated matter to calculate the value of electricity to a power system since that depends upon many factors including the particular plant mix of the individual power system itself. However, Electricity Boards in the UK are required to publish the rates at which they are prepared to buy privately generated electricity to cover their avoided costs and those figures therefore should include all the complexities. They are typically around 2.8 p/kWh for wind.

The margin between the generation costs of around 1.7 p/kWh and the value of the electricity to the power system of around 2.8 p/kWh is considerable. Wind energy's claim to be the cheapest method of generation available appears to be well substantiated.

7. CONCLUSIONS

Experience with modern wind farms such as the WEG one in California shows that the technology is now well established for medium sized wind turbines.

Based upon this practical experience, the economic case for wind energy on good, windy sites in Europe seems to be well substantiated.

The remaining problems are environmental and institutional and it is against this background of proven technology and economics that those problems need to be tackled.

8. REFERENCES

1. BWEA, Wind power for the UK, ISBN 1870064 01 1 (BWEA, London: 1987)

2. Lindley, D Economical windfarming Chartered Mech Eng Feb 1988 pp17-19.

3. Lindley, D and Quarton, D WEG's operational experience with two medium sized wind turbines Proc IEE, Vol 134, Pt A, No 5, May 1987, pp441-449.

4. Lindley, D and Quarton, D The design, construction and operation of three horizontal axis wind turbines for the generation of power Proc ICE, Vol 80, Pt 1, August 1988, pp969-997.

5. Lynette, R Wind power stations: 1985 performance and reliability, EPRI Final Report AP-4639, June 1986.

6. IEA, Renewable sources of energy p298 (OECD, Paris: 1987)

7. Swift-Hook, D T Decomissioning wind turbines pp37-38 Wind Energy Conversion, Ed J M Galt (MEP, London: 1987)

Chapter 2

The impact of environmental aspects on wind energy in The Netherlands

J. C. Berkhuizen and A. F. L. Slob

SYNOPSIS

The Research Department of Energy Unlimited ("Energie Anders Foundation") has completed a study on windenergy potential in the Netherlands taking into account environmental and physical planning aspects. In this study a model was developed which enables us to compute wind energy potential under pre-selected conditions such as constraints for noise, safety, electromagnetic interference, nature, landscape, etc. and possible siting of windturbines in areas where restrictive policies are in force.
In order to realize (large) windenergy projects it appears to be necessary to describe locations more in detail using guidelines to avoid environmental problems as much as possible. Special attention will be paid to safety and noise aspects, the influence of windturbines on birds and public and political-acceptance.

1. INTRODUCTION

In 1984 the Dutch Ministery of Housing, Physical Planning and the Environment wanted Energy Unlimited to answer the following two questions:
"What is the maximum wind energy potential in the Netherlands, when environmental and physical planning aspects are taken into account?" and
"What are the effects of the separate environmental and physical planning aspects on the wind energy potential that can be realized in the Netherlands?" (1)

In order to answer these questions, a computermodel was made. Its flexibility and the possibility of handling large amounts of data make the use of such a model quite effective. The computermodel consists of two parts:

- a database and
- a computing section

The database contains geographical information about physical planning and the environment, which is relevant to wind energy. The computing section is capable of handling information on different windturbines, different types of environmental constraints, different pre-selected economical factors and so on. It can then calculate the amount of the placable wind energy potential, where to place turbines and what the price per kWhour will be. Figure 1 shows where it ischeapest to place turbines with a rated output of 1000 MW with minor environmental problems. It appears that the windy areas with a small number of inhabitants in the northern part of Holland and the western part of Friesland have the best possibilities for wind energy.

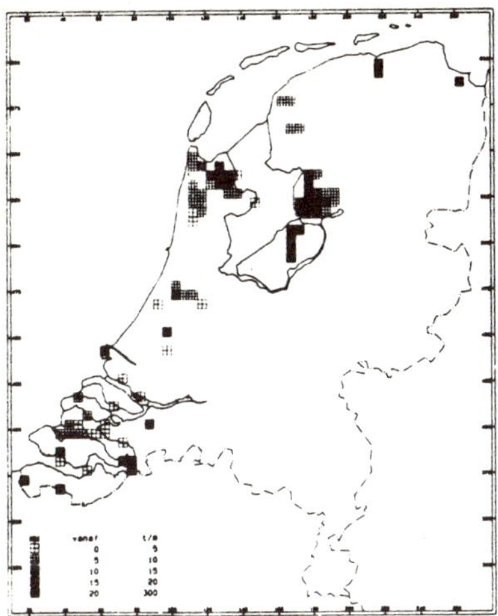

Figure 1: The cheapest 1000 MW in the Netherlands.

The model is also capable of calculating the theoretical influence of the different environmental aspects on the amount of wind energy potential.

Figure 2 shows that the nature aspect (including damage to birds), noise and landscape are major influences on wind-energy potential whereas telecommunication and safety aspects hardly influence the potential.

The model was finished in 1986. After that date about 15 MW of windpower has been realised in the Netherlands.

From the realization and its problems it became clear that in the local situation public and political acceptance is also a very important issue. Of these important aspects, a short description of the state of affairs in the Netherlands will be given below.

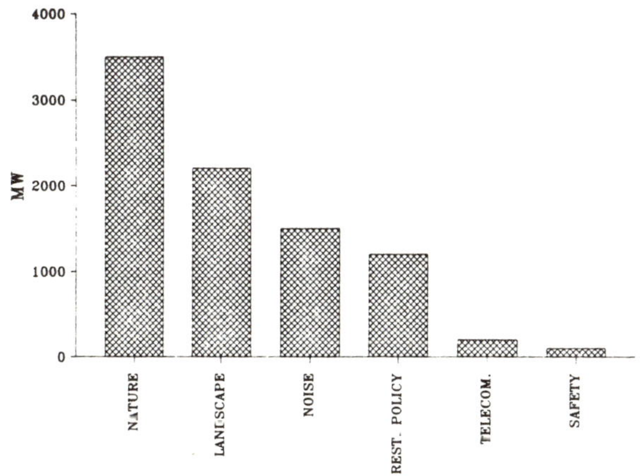

Figure 2: Theoretical influence of environmental aspects to the potential.

2. DAMAGE TO BIRDS

There are two possible effects that can damage birds:

1. birdvictims as the result of collisions against tower or blades
2. the disturbance of breeding or resting birds in the vicinity of the turbines.

The first effect has been and still is being researched at the Research Institute for Nature Management, at Energy Unlimited in the Netherlands and at the Game Biology Station in Denmark. In general, it can be said that so far results indicate that smaller turbines have no dramatic impact. This is justified by the small number of reported birds found dead near windturbines with heights less than 50 m. About the disturbance of birds much is still unknown because no research has been undertaken yet. At this moment more information is needed about the following situations:

- high windturbines and windfarms on locations near the coastline which are often mayor migrating routes
- windfarms in areas of periodically high densities of birds

This kind of research should ideally be undertaken on an international scale because the impact of windturbines on bird-life is of international interest (birds actually do not acknowledge national frontiers). The advantages of an international approach are that greater numbers and different types can be studied on a larger scale and that duplication can be prevented.

Wild-life-protecting organisations in the Netherlands closely follow the developments of wind energy for fear of damage to birds, specially when many windturbines, or turbines, that are very large (more than 50 m) are erected on locations where much bird-movement takes place. In the Netherlands many notices of objection have been brought in against wind energy-projects by conservationists, which have partly frustrated the introduction of wind energy. If no extra information on this aspect is provided the development of large turbines might be risky because of possible future siting difficulties.

On the short term this information won't be available. Therefore there must be an agreement on how to handle these uncertainties. In the Netherlands Energy Unlimited is now working on a so-called "social-contract" between wind energy developers and proprietors on the one hand and conservationists on the other. This document will have to contain a map indicating the areas where wind energy will be possible and where not, and areas where studies have shown possibilities to exist. Besides that map the contract will consist of an outline containing the mutual commitments, research needed and so on. This contract will place conservationists in a position to give directions to windenergy developers so as to avoid unacceptable risks. After evaluation wildlife-protection organizations will then have to tolerate further development if no significant negative results are found. Equally, wind energy developers must be able to put an end to experiments at risky sites if intolerable negative effects are found.

3. SAFETY

In contrast to other aspects of wind energy no research has taken place to assess safety levels in the vicinity of turbines in the Netherlands. There are extended requirements for windturbine construction as such and current policy is, that because of these requirements, windturbines should be safe. But remaining risks in the vicinity of a turbine have never been adequately assessed and it is unknown whether or not these risks are acceptable. In our opinion risks for people working, living or just being in the vicinity of windturbines should be tested by scientific methods (2).

According to the philosophy of other risk-analysis, safety levels can be assessed in terms of so-called individual risk, that is the chance that a person in the vicinity of an installation should die as a result of an accident. That level becomes higher when that person moves closer to a windturbine. In this way individual risk depends on the distance to the installation and can be shown in a so-called risk-distance-diagram.

In the Netherlands a political choice was made for an acceptable risk equal to or less than 10^{-7}. This means that the chance of death as a result of an accident is one to 10 million. Before such a diagram can be made for wind energy, more must be known about the frequency of failures.

Figure 3 shows an example of a risk-distance-diagram as the result of a blade-fracture from a large 3 MW-windturbine. For this example the frequency of blade-fracture is presumed to be once in a thousand years. The iso-risk-lines are calculated by a computer-simulation. In the blanc area housing is possible, whereas it is impossible in the double shaded zone. In the zone in between calculations should be made to assess actual safety-levels in order to compare this level to that of other places.

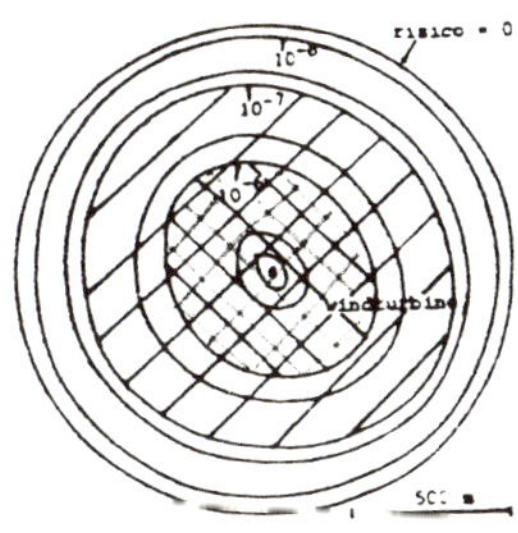

Figure 3: Risk distance diagram (after (3))

4. NOISE

The occurence of nuisance of noise, depends on two factors; first on the level of the acoustic emissions of the turbine and second on the distance between turbine and nearest residences. Or, in other words, on the emission level and the acceptable maximum immission level.

Acoustic emissions of windturbines can be divided into a mechanical component and an aero-dynamical component. Analysis of these sources has shown that for smaller turbines, with a rotor-diameter of up to about 20 m, the mechanical component is the most important factor, whereas for larger turbines, the aerodynamical component is recisive (4). If windturbines must be made more silent, efforts must be made in the area of this dominating component. At this moment a brochure is made in the Netherlands with rules of thumb for designers, on how to make individual windturbines as silent as possible.

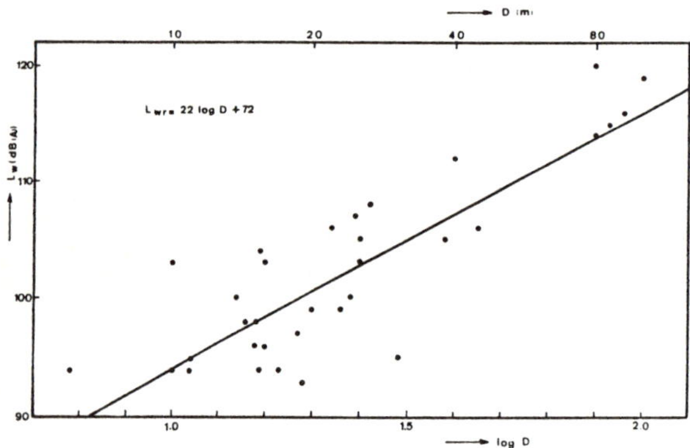

Figure 4: Measured acoustic emission levels (Lw) as function of rotor diameter (D) (4).

Figure 4 shows some measured acoustic emission levels dB(A)-corrected. The emission-level is shown as a function of the diameter of the turbine. There appears to be a relationship. In the last few years a reduction in emission levels of about 5 dB(A) has already been realized.

The second very important parameter in avoiding nuisance due to the acoustic emissions of windturbines is the assessment of an acceptable immision level. In the Netherlands, there is a special law to prevent nuissance of noise. This law pursues to keep silent areas as silent as they are according to the so-called "stand-still-princip". This means that requirements are very strict and permit a maximal immission noise level of 40 dB(A) near houses, at a windspeed of about 5-7 m/s. The windspeed is important, because both the emission-level of the turbine and the level of the background noise vary with the windspeed. A windspeed of 5-7 m/s has been chosen because in the Netherlands that windspeed is considered to be the windspeed at which the turbine noise is most distinctly audible.

When emission levels and the acceptable immission levels are known, it is easy to calculate the distance that has to be kept between windturbine and housing in order to avoid nuisance. Recent studies however have shown that there is an other important parameter influencing this distance and that is the number of turbines and distances between them.

Figure 5 shows how fast the necessary distances grow by increasing the number of turbines from one to ten turbines and a distance in between of a hundred metres.

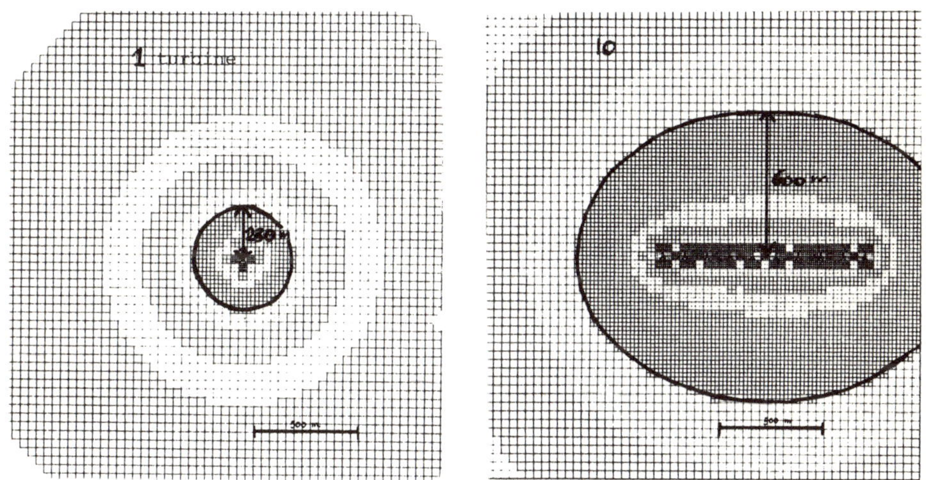

Figure 5: The distance of the 40 dB(A)-contour for one and ten windturbines (The black arrow shows the distance in metres of the 40 dB(A)-contour).

5. PUBLIC ACCEPTANCE

Studies have been carried out to find out what the attitude of the Dutch population is towards large scale wind energy projects. At this moment this attitude appears to be very positive. However, it has also been estimated that, in general, knowledge of wind energy is stil inadequate. This indicates that the positive attitude is rather unsteady and can easily be reversed.

A recently conducted study shows the surprising result that the possitive attitude of people, living near to a 1 MW windturbine shows a significant increase after the turbine has been erected (6). This is compared with the situation before the turbine came. People often seem to fear the coming of windturbines but once the turbine is operating the nuisance appears not to be as bad as they had expected. But, referring to the above mentioned need of a risk-analysis, a dramatic accident might throw reverse that positive attitude into a negative one. This will have its repercussions on, for example, the political need to introduce windpower on a large scale and the use of windpower as an option for the clean production of electricity. Trying to avoid a negative attitude towards wind energy also means that windturbines must be fitted into the landscape as well as possible. Landscape-architects have mentioned criteria to asses the possibilities of a good fit. Important in this respect are land-use, the visual structure, orientation and history, and the visibility and the size of the turbines versus the scale of the landscape.

In the Netherlands Energy Unlimited has made a list of infrastructures and elements, which might conflict with wind energy. The infrastructures and elements are roads, railroads, dikes, power lines, system of pipes, micro-waves, transmitters, radar navigation systems, military practice grounds, air ports, and nature reserves. These guidelines are mainly meant as a useful tool for sieve-analysis in order to find suitable locations for windturbines (7).

6. CONCLUSIONS

In the Netherlands many environmental aspects have been studied. There remain nevertheless some gaps in our knowledge. Going back to to Figure 2, showing the theoretical impact of environmental aspects on the potential of wind energy, it becomes clear that some aspects need more attention than others, which not only depends on their magnitude, but also on the remaining uncertainties. It becomes clear that the nature-aspect still needs much attention because the magnitude of the effect is rather large and the uncertainties too. The same can be said about the magnitude of noise, but here the uncertainties are rather fewer, meaning that it is easier to take their impact into account; the impacts are predictable. For the safety-aspect the opposite is true; the magnitude might be small, but the uncertainties are enormous.

The following conclusions then can be drawn:

1. The importance of environmental aspects depends on the magnitude and the uncertainties of the effect,

2. The most important environmental aspects at this moment are noise, damage to birds and safety-levels in the vicinity of a turbine.
3. In the Netherlands it is now possible to deal with these aspects in the short term, because of the many conducted studies, but more information is needed before a very large number of windturbines can be sited.

7. REFERENCES

(1) Berkhuizen, J.C., J.C. van den doel, A.F.L. Slob and E.T. de Vries, 1986: "Estimation of the Wind Energy Potential in the Netherlands, taking into account environmental aspects" EWEC '86 proceedings, Rome.
(2) Vries, E.T. de, in prep: "Omgevingsveiligheid van windturbines", CEA, Rotterdam.
(3) Montgomery, B., 1982: "Horizontal axis windturbineblade failure," Sweden.
(4) Borg, N.J.C.M. van der, W.J. Stam, W.B. de Wolf, 1988: "De akoestische bronsterkten van windturbines", ECN-88-039, Petten.
(5) Bosma, S., A.T.H. Pruyn, G.J. Kok and F.W. Siero, 1985: "Houdingen en weerstanden ten aanzien van klein- en grootschalige windenergie", WEC '85, CEA, Rotterdam.
(6) Wolsink, M., 1986: "Public acceptance of large WECS in the Netherlands", EWEC '86 proceedings, Rome.
(7) Berkhuizen, J.C., E.T. de Vries and A.F.L. Slob, 1988: "Siting procedure for large windenergy projects" Journal of Wind Engineering and Industrial Aerodynamics, 27 (1988): 191-198.

Chapter 3

Legal aspects of wind energy exploitation: an EEC perspective

N. O'Neill

SYNOPSIS This paper examines a range of legal issues, particularly in the environmental area, which affect the exploitation of wind energy. Existing or proposed European Community legislation on, inter-alia, environmental impact assessment, technical standards, product safety, state aids, and the free movement of goods will be reviewed to the extent that they concern wind energy systems. Steps are suggested which might enable the legal environment, as much as the natural environment, to promote rather than hinder the exploitation of wind energy.

1. INTRODUCTION

Those responsible at European Community level for the formulation and implementation of EC energy policy have already noted that the development of new and renewable energy sources can be restricted by a range of non-technical barriers. This concern was expressed in 1986, in the Resolution adopted by EC Energy Ministers (1) which called for action to "ensure cooperation at Community level with a view to the coherence, if necessary, of national legislative, financial and information measures." The same Resolution went on to call on Member States "to examine the need to set up agencies ... to promote the use of new and renewable energy sources in order to advise contracting public authorities, local authorities and small and medium sized businesses in the planning of feasibility studies and on the technical and financial aspects of implementing projects to exploit these sources;".

This concern over measures which could hinder the development of new and renewable energy sources was again evident in the EC Commission's Communication (2) of September 1987 on developing the exploitation of renewable energy sources in the Community. The Communication states that " The main obstacles, apart from further technological development needed for some applications, are the lack of knowledge among decision-makers of what has been demonstrated and what is now viable, the absence of tariffs likely to attract renewable-energy-generated electricity sales to the public grid, and the need to establish common principles for the certification of equipment so that there is no hindrance to the internal market in renewable energy equipment. In addition, administrative procedures need to be streamlined." The Communication goes on to point out that for all renewable energies action is need on:

- clear and expeditious authorization procedures, with due regard for environmental aspects, by means of straightforward and fast decision-making processes which can only be achieved though specific provisions for renewable energies, still lacking in several Member States.

- the free movement of exploitation equipment in the Community and the introduction of common certification principles aiming at a more rational and wider market.

- contractual relations with the public producers/distributors whose grid takes part or all of the electricity generated by renewables.

Specifically on wind energy, the Communication states that "the main problems (for).. rapid development .. lie in the difficulty of obtaining construction permits for many sites, mainly for environmental reasons, and in the existence of too many manufacturers of wind turbines in the Community, given the likely market. Furthermore the obligation in some Member States to certify wind turbines at present limits intra-Community trade." A closer look at the dual problems of construction permits and technical standards is therefore justified.

2. PLANNING LEGISLATION/CONSTRUCTION PERMITS

Wind energy systems like other industrial installations require a building or construction permit before they can go ahead. The granting of such a permit is usually dependent on the system being in compliance with the relevant physical planning legislation. Physical planning is an area in which the the EC has no legal competence, the relevant powers being reserved by the Member States. They in turn have generally delegated responsibility for physical planning to regional and/or local authorities often only maintaining for themselves the right to adopt framework legislation.

Perhaps the greatest difficulty wind energy systems face with planning legislation is that it was never formulated with wind energy in mind, and that once established such legislation has a sort of in-built aversion to change. The Community's experience of the problems which the exploitation of wind energy has had with planning legislation is primarily drawn from Denmark and the Netherlands. In Denmark, difficulties in obtaining construction permits have until now generally been few and far between, aided by the low population density, the support of local authorities for wind energy systems and a constructive attitude on the part of the national authorities. But now there is alas "something rotten in the state of Denmark". Last Summer several Danish organisations called for a review of the authorization process for wind turbines. They called for a ban on the siting of wind turbines in EEC designated bird protection areas, scientific reserves and nature reserves. Moreover, turbines should be prohibited from within 300 meters of certain buildings and, if approved, be more often subject to what amounts to an environmental impact assessment procedure. Denmark is, however, a rather special case given the very high number of individual wind turbines which have been installed. Elsewhere in the Community, there will probably be a tendency to install wind parks with an average installed capacity greater than that seen in Denmark.

To-date, it is in the Netherlands that the greatest difficulties with planning legislation have been encountered. This is perhaps not surprising given that many of best wind energy sites are in areas with limited available land resources. Moreover, physical planning provisions are quite extensive and an often time consuming procedure must be followed in order to obtain all the requisite authorizations. Nevertheless, the Dutch have made considerable efforts to overcome physical planning restrictions. As far back as 1983, the Association of Dutch Local Authorities produced an information booklet on wind turbines for their members. ore recently, the Ministry for Economic Affairs under whose auspices the wind energy policy falls, has acknowledged the need for further information to be given to local and provincial authorities so that they can better assess planning applications for wind energy systems. At the same time, part of the wind energy research budget will be allocated to studies on the administrative, planning and societal aspects of the greater use of wind energy. More interestingly, as part of a better information policy, the Minister for Economic Affairs together with the Minister for the Environment will address to all local authorities, in good wind regions,a circular on improving the procedures for planning applications for wind energy systems.

In addition, the Ministry for the Environment in cooperation with local and provincial authorities has designated 25 environmental zones suitable for the siting of wind energy systems. Turbines sited in these regions are also eligible for an additional government subsidy. Furthermore, research already undertaken has convinced a number of provinces to alter their planning requirements to permit the placing of individual wind turbines with a hub height of up to 25 meters (wind turbines nearly always fall foul of existing planning requirements as the intended hub height exceeds the maximum permitted within the local authority area). The Dutch government has also accepted the fact that present planning legislation would only enable 600 MW of wind energy to be installed by the year 2000 rather than the 1000 MW which is the government's stated policy objective. In order for the objective to be attained it has been proposed that the government should introduce guidelines on the siting of wind turbines, particularly alongside stretches of open water, and also reserve sites suitable for the installation of large scale wind turbines.

The Dutch policy on planning legislation is interesting insofar as it acknowledges the importance of information in overcoming difficulties in obtaining building permits. Moreover, the deliberate cooperation between the Economic Affairs/Energy Ministry and the Environment Ministry is an important lesson which can be followed in other EC Member States. In most Member States the exploitation of wind energy continues to come within the scope of Energy Ministries with limited input from Environment Ministry colleagues. Given the environmental/physical planning aspects of wind energy turbine siting there is a lot to be said for inter-Ministerial cooperation.

Other EC Member States might well consider the advantages to be gained from a greater information campaign vis à vis local authorities. Such a campaign could request those local authorities in areas particularly suited for wind energy exploitation to adopt no new zoning plans which might hinder the siting of wind energy turbines and indeed to undertake the preemptive revision of existing plans. In addition, Government Ministries directly responsible for the administration of large land areas e.g. Forestry Commissions, Defense Ministries, Environment Ministries ought to revise their own planning requirements to remove potential barriers to the exploitation of wind energy, particularly if its development and expansion is part of the government's energy policy objectives.

Apart from a general incompatability between existing planning requirements and the installation of wind turbines, as well as the lack of information available to local decision-makers, additional difficulties on the siting of wind turbines can arise over concern about the general environmental impact of a turbine. This concern can focus on the aesthetic nature of the turbine, noise emission, danger to the bird population, and electromagnetic interference. Some of these particular issues will be addressed in other papers. The global analysis of the abovementioned questions can , however, be undertaken by an environmental impact assessment. To-date, few wind turbines have had to undergo the full rigours of such an assessment, but this may now change.

In June 1985, the EC adopted the environmental impact assessment (EIA) Directive (3) which should be implemented by all Member States by 3 July 1988, at the latest. The Directive applies to "the assessment of the environmental effects of those public and private projects which are likely to have significant effects on the environment". For certain projects an EIA will be obligatory whereas for others it can be optional. The Directive states that " Projects of the classes listed in Annex II (of the Directive) shall be made subject to an assessment,... where Member States consider that their characteristics so require". Among the projects listed in Annex II of the Directive are " Industrial installations for the production of electricity, steam and hot water (unless in included in Annex I) ." Annex I refers to thermal power stations with a heat output of 300 MW or more and to nuclear stations.

At the end of January 1988, the UK Department of the Environment issued a draft circular on the implementation of the EC Directive and invited comments thereon (4). For the purpose of identifying those projects in Annex II which ought to be subject to an EIA the circular states that " The basic question to be asked is whether a project is likely to give rise to significant environmental effects". The circular goes on to recall the aspects of the environment likely to be significantly affected by the proposed project (Annex III of the Directive). These cover " population, fauna, flora, soil, water, air, climatic factors, material assets, including the architectural and archaeological heritage, landscape and the inter-relationship between the above factors".

The circular indicates that in general terms the Secretary of State feels that an EIA will only be necessary for:

- those really major projects which are of more than local importance;
- a few smaller projects proposed for particularly sensitive locations;
- in exceptional cases, for projects with unusually complex and potentially adverse environmental effects, where expert and detailed analysis of those effects would be desirable and would not be carried out under normal planning procedures.

On the basis of the above criteria, the Secretary of State has suggested that no more that a few dozen Annex II projects would be subject to an EIA each year. For wind turbines, the question of siting in particularly sensitive areas will be of importance. Again according to the draft circular " Consideration should be given to the need for an EIA where an Annex II project is likely to have significant effects on the special character of a protected area or site, such as a national park, a site of special scientific interest, a national nature reserve or an area of archaeological importance. Any views expressed by the Nature Conservatory Council, the Countryside Commission or the Historical Buildings and Monuments Commission should be taken into account. Authorities may wish to consult those bodies where they are in any doubt about the significance of a project's likely effects on the natural or built heritage".

Finally, the circular stages that special considerations will apply to special protected areas under EC bird protection legislation. In such areas there will be a particular obligation to avoid developments which could have damaging effects on the environment and any such developments should be subjected to the most rigorous examination.

In an appendix to the circular, the Department sets out a series on indicative criteria and thresholds for the identification of Annex II projects requiring an EIA. None of these criteria refer to energy projects. A second appendix to the circular sets out the procedures for identifying projects requiring an EIA. In relation to Annex II projects, it states that the planning authority are on receipt of an application for a project obliged to consider whether in their opinion the characteristics of the development are such that it is likely to have significant environmental effects. If so, they must notify the developer within three weeks that planning permission ought not to be granted without consideration of an EIA. Finally, as regards projects submitted by the CEGB or Area Boards in respect to " smaller power stations "the Secretary of State for Energy will be given a discretionary power on whether to require an EIA.

To sum up, an EIA is not specifically foreseen in the case of wind turbine installations but the likelihood exists that it may be sought for siting in particular locations and perhaps for any wind parks. So far an EIA has only been undertaken for the Sunsetter Hill turbine on the Shetlands. In the event that EIAs are required for further developments they will probably have to be more extensive than the Sunsetter Hill EIA. More extensive also means more costly. The wind energy industry must also be aware of how other EC Member States classify wind energy developments under the EIA Directive.

3. TECHNICAL STANDARDS

For the past number of years organisations in various Member States have been working on both the drafting of technical standards specifically for wind turbines and on the establishment of certification procedures for wind energy systems. Specific standards for wind turbines are probably most advanced in the Netherlands where NEN Norm 6096 applies. Relevant technical standards have, however, also been identified in Denmark, Germany and the United Kingdom although not always specifically for wind turbines. Certification procedures are most advanced in the Netherlands and Denmark mainly on account of subsidies only being available to certified turbines. The certification of turbines fulfills two objectives: offering greater certainty to the purchaser as regards performance and safety, and also reassuring planning authorities as regards safety. The present difficulty, however, lies in the lack of common technical standards or certification procedures within the EC. This results in the unfortunate spectacle of a certified Danish turbine having to undergo new tests if it wishes to be installed in the Netherlands or Germany and vice-versa. This difficulty could be resolved in a number of ways, most notably by the establishment of common standards and certification procedures throughout the Community or the mutual recognition of prevailing national standards and procedures.

The EC Commission for its part has tried to resolve the question of technical standards by supporting cooperation between wind turbine test centers in the Netherlands, Germany, France, Denmark and the UK. The objective being to create commonly accepted standards or at least the basis for establishing the equivalence of national standards which would allow for mutual recognition. Two recent legal developments have given an added impetus to the search for common technical standards. Firstly, there is the Community's objective of completing the Internal Market by 1992 which is set out in the Single European Act. In order to promote the free movement of goods, the EC Commission has undertaken to present a variety of proposals to remove existing trade barriers.

In the event that Community legislation does not remove a particular barrier by 1992, article 100b of the EEC Treaty now provides that " During 1992, the Commission shall, together with each Member State, draw up an inventory of national laws, regulations and administrative provisions which falls under Article 100a and have not been harmonized pursuant to that Article. The Council acting in accordance with the provisions of Article 100a, may decide that the provisions in force in a Member State must be recognised as being equivalent to those applied by another Member State."

Thus if some common standards have not been agreed by 1992 there will be a certain onus on the Commission to end discrepancies existing at that date. More important, however, is a recent EC Commission proposal for a Directive on machine safety (5). This proposal is based on the Community's so-called new approach to technical harmonisation. The new approach basically means that the EC Commission will no longer seek total harmonization as regards product specifications but will instead confine itself to setting essential safety requirements at EC level. Once a product is capable of meeting the essential safety requirements and is so certified it should be able to freely circulate throughout the Community. The technical standards which manufacturers should follow will not been established by the EC Commission as such but instead their preparation is contracted out to recognised standardization bodies such as CEN/CENELEC. The standards would not, however, be compulsory, a manufacturer could choose to use other standards but would have to show that the use of such standards still allows him to respect the essential safety requirements as set out in the relevant Community Directive. For series production, conformity with the essential safety requirements would be on the basis of self certification.

As it presently stands, the proposed Directive on machine safety would apply to wind energy systems. The essential safety requirements as set out in the annex to the proposal would, amongst other things, cover the principles of safety integration, the materials and products used, control devices, stability, risk of breakup during operation, protection against hazards due to moving parts, fire, noise, vibration and information provided to the user. Although this proposal has not been drafted by DG XVII of the Commission it is to be hoped that when mandates are granted for the preparation of standards to meet the essential requirements, the work could be undertaken by the group presently being sponsored by DG XVII. In any event it looks most likely that the day of common standards for wind energy turbines within the Community is coming closer to hand.

Until such common standards prevail certain difficulties still remain. As regards EC legal remedies, all Member states are presently required under the provisions of EC Directive 83/189 to notify the EC Commission of all new draft technical standards. Notification of the draft automatically suspends its implementation for three months, objections by other EC Member States for six months, and if the Commission is itself planning legislation in the area the suspension of the draft standard's implementation can be extended to a period of one year. Of course, should the EC Commission take the view that the draft standard represents a barrier to trade within the Community it can seek to have the standard amended and, failing that, institute proceedings against the Member state concerned before the EC Court of Justice.

With regard to technical standards or national legislation presently in force and which are thought to hinder free trade (Danish wind turbine manufacturers allegations on German requirements come to mind), the option already exists to complain to the EC Commission. If the complainant can convince the Commission of the legal justification of the case, the Commission can take action against the offending Member State, moreover such action is at the expense of the Commission and not the complainant. Unfortunately, legal proceedings rarely have the effect of suspending application of the contested measure. Finally, one should note that at the end of this year the EC Commission has undertaken to convene a conference on the barriers to the greater exploitation of renewable energies.

The conference should bring together public organisations and authorities, manufacturers, financial bodies, and scientific and technical institutes with the aim of finding solutions to a variety of technical administrative and legal barriers. The conference will be an important opportunity for the wind energy industry to voice their concerns on a number of issues.

4. PRODUCT LIABILITY

The need for manufacturers to produce safe wind turbines is self evident, but no wind turbine, like any other product, can be guaranteed to remain safe throughout its working life. Most manufacturers are willing to guarantee the safety of a wind turbine on delivery and indeed, if not legally obliged to do so in all EC Member States at present, will be so required within the next few years. This need to meet a general safety requirement should not prove to be unduly difficult. The main problem may arise when a product already on the market develops a fault. If such a fault merely affects the machine's performance it can probably be rectified under guarantee provisions, maintenance contract or if no satisfactory solution is found by recourse to litigation for contractual liability. Such litigation has recently become a feature of the Californian wind energy industry but is so far generally absent in Europe.

In the situation where a fault in a wind turbine actually causes personal damage one is no longer talking about contractual liability but rather product liability. The laws within the European Community on product liability are presently undergoing change following upon the adoption of an EC Directive on product liability in 1985 which should enter into force in all Member States by the end of July of this year. So far its provisions only apply in the UK under the guise of the Consumer Protection Act, Part 1. The main impact of the new product liability legislation is that it introduces throughout the EEC the concept of strict liability. This basically means that the person injured no longer has to show that the fault in the product was due to negligence on the part of the producer (or operator) but only that a fault did actually exist and that the fault caused the injury involved.

From the viewpoint of wind turbine manufacturers, the advent of strict liability regime is important insofar as the industry is still to a certain extent in a development phase, and thus the risk of accidents with equipment is somewhat greater. In the event of a wind turbine manufacturer being sued for damages under the product liability Directive,one defence which may be available is the "development risk defence". At its simplest this would permit a manufacturer to avoid liability if it could be shown that the state of scientific and technical knowledge at the time of the machine's production was not such that the manufacturer could have discovered the fault. The defence may not, however, be available to manufacturers in all EC Member States as it is an optional provision in the Directive. As of March 1988 it appears that France, Belgium and Luxembourg may not allow for the defence.

As a precaution against product liability actions, turbine manufacturers should be careful as to claims made about their machines so as not to create a false and dangerous sense of security and unrealistic expectations as regards safety. If the actual design does not meet the expectation created in consumers via advertising, the producer may be held liable for any damages resulting from this. In addition, the lack of adequate warnings or operating instructions may render a product defective, as could instructions for repair and maintenance. Thus care should be taken with regard to the readability, comprehensibility and language of all instruction manuals. Moreover, it should be ensured that the operator has received a manual. Manufacturers would also be well advised to keep good records on quality control tests, dates of sale and servicing and of any complaints received and follow-up action. A system of informing customers with a particular model of any potential safety problems should also be provided for. Finally, the manufacturers' compliance with existing technical standards will assist in any defence against a product liability claim but will not be sufficient to enable a denial of responsibility.

5. STATE AIDS

The exploitation of wind energy in most EC Member States has been assisted by government finance especially for R&D work. Moreover, in the Netherlands and Denmark subsidies have been available for the actual purchase of wind turbines. Any funds allocated by the Member States for R&D must be in the context of an aid programme which has been approved by the EC Commission(6), otherwise firms benefitting from the aid could find themselves being required to pay it back. Moreover, if manufacturers in one Member State feel that they are being placed at a competitive disadvantage by the state aid granted to their competitors in another Member State they have the right to request the EC Commission to investigate such an aid programme. As the market for wind energy systems develops in Europe and elsewhere, the Commission will undoubtedly take a closer look at the justification for state aids, especially if there appears to be overcapacity in the industry.

6. ELECTRICITY GENERATION

The legal situation regarding an individual's or company's right to generate electricity is far from uniform in the Community. A similar situation applies to the right to sell electricity either to the national grid, local distribution companies and eventually to third parties. There have been developments in several Member States over the past number of years e.g. UK, the Netherlands, Denmark and Greece. Nevertheless, the case remains for the establishment of a possible model contract between wind energy producers and electricity companies. While not fixing the actual prices to be paid to turbine operators, a model contract could establish a set of similar operating conditions. Moreover, it might also address the issues of the cooperative ownership of wind turbines and the requirement for the owner to live within a certain distance of the turbine. There could well be a coordinating role for the Commission in such a project.

7. INTERNATIONAL TRADE

As the market for wind energy systems develops so too will competition in the market place. The European market has so far been dominated by EC firms, as has the Californian market. Elsewhere in the world the presence of Japanese manufacturers has increased. As non-EC producers increase in strength, it is well for Community manufacturers to remember that EC Community legislation exists to protect them from economic injury on their home market from dumped foreign turbines. Illicit trade practices by third country governments in non-EC markets can also be tackled.Finally, unjust tariff barriers will not be removed by national administrations but rather by Community action.

8. CONCLUSIONS

In the environmental area the main difficulty for expanding the exploitation of wind energy is the difficulty in obtaining planning permission. The Dutch policy of a coordinated Ministerial approach to local authorities coupled with providing the maximum of relevant information appears to be a useful approach. Similarly the designation of areas particularly suited to wind energy is to be recommended. A variation on this could be the designation of wind energy as the preferred energy source in certain areas, perhaps on offshore islands and in isolated areas. The implementation of the EC environmental impact assessment Directive by the Member States should be followed carefully so as to ensure that it does not lead to new restrictions and possibly place some manufacturers at a competitive disadvantage.

Progress towards the realisation of common technical standards and certification procedures will increase in the coming years. The EC Commission's undertaking to more closely examine non-technical barriers to the exploitation of renewable energies should be fully availed of. Increased technical certification will also enable wind turbine manufacturers to more easily observe new EC provisions on product safety.

REFERENCES

(1) Council Resolution of 26 November 1986 OJEC No. 316, 9/12/1986.
(2) COM (87) 432 final of 29 September 1987.
(3) Council Directive of 27 June 1985 on the assessment of the effects of certain public and private projects on the environment. OJEC No. L173 5/7/1985.
(4) Environmental assessment : implementation of EC Directive, Department of the Environment 26 January 1987.
(5) COM (87) 564 final of 14 December 1987.
(6) Community Framework for State Aids for research and development. OJEC No. C. 83,11/4/ 1986.

Chapter 4

Local authority rating of the WEG Ltd. wind turbine generator in Devon

R. Lord

Introduction

The WEG prototype 25 m diameter, 250 kW wind turbine generator, was installed in Ilfracombe, Devon at the end of 1984 and has remained on the site since then.

Local Authority Rates are the local taxation in the UK from which revenue the local authorities fund the majority of their expenditure on such items as education, police, fire services, refuse disposal and street lighting. In recent times it has been based upon the assessed rental income which could be gained from a hereditament (property) if it were rented to a third party. Where the rental value of a hereditament can not be asessed, the law provides methods of calculation which are aimed at providing an equivalent rental value. Taxes are then paid to the local authority, based on these rateable values. Unfortunately wind turbine generators are generally subject to these taxes.

This paper aims to give a summary of the author's recollections of the history of the Rating for tax purposes of the WEG Ltd wind turbine generator in Devon.

Early Days

In the early nineteen eighties, the BWEA became interested in the subject of rating of wind turbines and took advice from chartered surveyors expert in rating matters. This was somewhat inconclusive and so meetings were sought with the Department of Environment, which is responsible overall for local authorities. A Queen's Council was also asked for his advice on whether wind turbines were rateable (ref Appendix 1), and if so what the value of the rating would be.

No firm conclusion was reached from all these learned discussions, though all agreed that wind turbines would be rated. They thought that of the 3 main methods of evaluation:-

1. Direct evidence of what a hypothetical tenant would pay to rent the hereditament.
2. The profits test, where some percentage of gross receipts is taken as the rental value (can only be applied where there is evidence from similar existing examples).

3. The Contractors test, which is based upon 6% of the capital cost of the hereditament in terms of 1973 monies.

that the last method was the most likely to be used.

The English system is based on precedent and case law, so the only real way to find out was to build a wind turbine; the Local Authority would then rate it, and the owner could then challenge the rating through the courts.

So the matters rested until the WEG wind turbine came up for rating.

Assessment of the Rates

The Inland Revenue is responsible for the assessment of Rateable Values, I suppose to ensure that they are uniform over the country, and then the tax is collected by the Local Authority.

On the 27 January 1986, the Inland Revenue Valuation Office requested details of costs relating to the construction of the wind turbine including its housing and site works.

Some discussion with the Valuation Officer took place, where the argument was put that as this was a prototype, costing was difficult and costs unrepresentative of current values. Finally, a fair basis was agreed as being the cost of one of a batch of production machines, plus the site civil engineering costs incurred. Hence the non-recurring costs of protoype development were ignored and a figure of approximately £120,000 accepted.

Then on 24 March 1986, WEG received the assessment of rateable value at £1638. For the year commencing April 1986, the Local Authority Rate payable was £2.0075 for every £1 of rateable value, so the tax due was £3288.29 for the year.

The first reaction was to give notice that WEG was unhappy with this valuation, and proposed to appeal. The second was to argue over the start date. Here WEG was able to convice the Local Authority that it should only pay rates from the end of the commissioning period - late in 1985.

Generating Authority Rates

Certain Authorities could not be rated sensibly under the normal rating evaluation system, and among these is the Central Electricity Generating Board, where it pays rates calculated by formula from its overall generation of electricity. Under this formula it pays of the order of .1p per kWh supplied to consumers.

The Local Appeal

The first appeal on rating matters is dealt with by the local Valuation Court where a decision is given by an independent lay tribunal, experienced in rating matters.

Discussion of the valuation with the Valuation Officer, revealed that the contractor's test had in fact been used in arriving at the valuation. This had been based upon the cost figures provided by WEG and although arguments could be presented for minor changes, was felt basically to be a reasonable application of the method (ref Appendix 2). It was the method of valuation that WEG felt was wrong.

The date for the appeal was fixed for 17 October 1986.

WEG's arguments were on several grounds:-

1. Planning permission had only been granted for 3 years and the wind turbine might have to be removed thereafter, thus reducing its potential rental value.

2. The local Environmental Health Dept had requested that the machine was not operated for a 10 hour period overnight, due to its proximity to local residences, again reducing its potential rental value.

3. The contractors test was unsuitable as a means of valuing wind turbines, as it did not give a fair valuation. For instance, a potential electricity generator could buy a diesel installation capable of producing the same electricity for about one fifteenth of the cost of the equivalent wind turbine. Both would have the same income from sale of electricity, but the diesel generator would only pay one fifteenth of the rates.

4. The electricity utilities sold electricity at typically 5p per kWh and paid .1p per kWh in rates while WEG with its wind turbine was forced to sell to a monopoly purchaser, the Area Electricity Board, at about 2.3 p per kWh while it had to pay 1.2p per kWh in rates. This was hardly equitable and also showed the Contractors test to be inapproprate.

The final argument was that this wind turbine should be formula rated as were the utilities, and that the Secretary of State for the Environment had been given the power to allow this under clause 34a of the 1983 Rating Act.

The Valuation Officer argued that the Contractors test was appropriate and had been correctly applied by him.

The Local Valuation Court ruled on 20 November, that the rateable value should remain at £1638 under current legislation.

Further Appeals

The Appellant has the right within 28 days to lodge a futher Appeal with the Lands Tribunal.

This was done immediately. However further consideration of the case led to the conclusion that current legislation was being correctly interpreted and so further appeals would only incur nugatory expense. The appeal was withdrawn.

Alternative Courses of Action

With the failure of the most logical course of action the alternative route was to persuade the Department of Environment that their Secretary of State should enact the powers he had been given by Parliament to formula rate wind turbine generators on a similar basis to the utilities.

WEG had a great deal of support in this activity, and highlighted below are some of the key players:

1. The Dept of Energy as the high rate burden prevented the Energy Act from being effective, and discourgaed private generators.
2. The Dept of Trade and Industry as it wished to encourage the renewable energy equipment manufacturers and hence create a new industry, and possibly exports from it.
3. The Dept of Employment as it wished to create employment in new industries.
4. The BWEA and Tradewinds.
5. Other renewable energy manufacturers including mini-hydro (AUR Hydropower) and also combined heat and power producers.
6. Members of Parliament where appropriate questions were asked both houses.
7. The Engineering Institutions as they wish to maximise opportunities for employment of engineers.

WEG itself has an enormous file of correspondence lobbying for changes in the rating system.

Other Rating Aspects

Under the UK system, a wind turbine on agricultural land producing power solely for agricultural purposes would not be rated-a curious anomoly of the system.

The WEG wind turbine also has to contribute to water management and drainage of the area based on its rateable value.

In the WEG windfarm in California, property taxes are payable based on the initial capital values, reducing over the first 5 years as the machines are depreciated. This Local tax is equivalent in the first year to .3p per kWh falling to about half this over 5 years.

If a wind turbine is owned by a utility it is rated under their formula and so pays about one tenth of that paid if it was owned by a private generator.

The Future

In 1990 the rating system is due for revision in England and Wales, and is due to be based on a pole count. That is for domestic rates, the property element disappears and the tax is to be paid on an individual basis. Industry is still to be taxed and working groups have been set up to examine ways in which industry can be taxed on an equitable basis. To some extent WEG's activities must have been successful as it has been invited to be represented on the working group which is to advise the Secretary of State for the Environment as to the rateable values to be applied to the electricity supply industry from 1 April 1990.

The terms of reference of the working group require it to relate rateable value to the industry turnover or added value by both a historical approach and a proxy method. The historical approach will consider revising the formula to give the industry the same tax burden as it had at the time of the last formula revision, taking into account changes in the business since then.

The proxy method will consider the relationships currently used to relate the property and assets used by industry generally to their rateable values and to determine whether this is an appropriate method for rating the electricity supply industry.

The working group is to receive some guidance from the Formula Rating Steering Group and has been asked to consider particularly:

- changes in business with time
- future adjustments to the formula
- sharing of rateable value between England and Wales
- application of the formula to other generators of electricity
- applicability of the formula in event of structural changes within the industry

The working group seems to have quite a task, but at least it is instructed to consider the fair rating of other generators of electricity. There is progress at last, but not for enactment before April 1990.

There is also a similar working group in Scotland where the wind industry is represented.

Conclusions

This story seems to be a triumph of bureaucracy over common sense, but at least there is light at the end of the tunnel - in 1990. The present local taxation system seems to prevent the generation of electricity from renewable energy sources by private generators from being viable, though there are opportunities for the generating utilities.

I would be pleased to learn of the situation in other countries to see if there are ways in which the UK system could benefit from experience overseas.

The wind industry should have an exciting future for private generators after 1990, with a more sensitive local tax regime and the privatisation of the supply industry.

BWEA Requests Legal Advice on Wind Turbine Rating Valuations

The Council of the BWEA has become concerned that rating valuation of wind turbines erected within the United Kingdom by other than the electricity supply industry could result in substantial local taxes commensurate with the value of the electrical output (wind turbines erected solely to supply power to farms would, however most probably be exempt). It was therefore decided to seek legal opinion which has now been obtained and is reproduced below for the information of members.

The Council has also been advised in this matter by a firm of chartered surveyors familiar with rating valuation in England, Wales and Scotland. As a result of this professional advice, representations have been made to the Secretaries of State at both the Departments of Environment and Energy to obtain a more realistic level of valuation for wind turbines.

Advice relating to wind turbine rating as received by the BWEA from counsel, Mr Guy Roots:

"I am asked to advise whether and to what extent wind turbines are rateable. A wind turbine comprises a building or structure which houses or supports plant and machinery designed to generate electricity by use of wind power. There are a variety of designs of wind turbine, but for present purposes precise details are not necessary.

The occupier of any land is liable to be rated (section 16 of the General Rate Act 1967). Every unit of property liable to be rated, known as an hereditament (section 115), must be entered in the Valuation List (section 67). An occupier's annual liability to rates is calculated by applying the rate determined annually by the rating authority to the rateable value of the hereditament entered in the valuation list (sections 2 and 3).

There can be no doubt that occupation of a piece of land for the purposes of constructing and operating a wind turbine used for the generation of electricity would give rise to liability to be rated. The extent of such liability is determined by the assessment of the rateable value.

Clearly the land occupied by and in conjunction with the wind turbine, together with the structure which houses or supports it, would fall to be included in the assessment. Apart from the structure, most if not all of the wind turbine consists, as I understand it, of plant and machinery. The extent to which plant and machinery is liable to be rated is governed by section 21 of the General Rate Act 1967 and the Plant and Machinery (Rating) Order 1960 (as amended).

In my opinion a wind turbine would fall within the term "windmill" which is found in Table 1A of Class 1A of the Order, and it would be "used or intended to be used mainly or exclusively in connection with the generation .... primary transformation or main transmission of power in or on the hereditament" (Class 1A). Accordingly, in my view, the plant and machinery of a wind turbine would fall to be assessed "up to and including: .....in the case of electrical power, the first transformer in any circuit, or where the first transformer preceds any distribution board or there is no transformer the first distribution board". It may be, therefore, that a relatively small amount of cable and equipment would not fall to be assessed.

The views I have expressed above relate to the law in England and Wales. In Scotland the legislation is somewhat different, but it is my understanding that wind turbines would be rateable to a substantially similar extent. However, I do not practice in Scotland and if the Association requires to be certain about the position there, separate advice should be sought"

SCHEDULE A

VALUATION OF WIND GENERATOR SLADE ILFRACOMBE

Effective capital value based on costings provided by Wind Energy Group as at 1985.

Rotor		£23,100		
Nacelle		£58,800		
		£81,900	Factor to reflect costings as at 1973	
			X 0.22	£18,018
Tower		£10,000		
Assembly		£10,000		
Site Works		£18,409		
		£38,409	Factor to reflect costings as at 1973	
			X 0.34	£13,059
Fees	£18,018			
	£13,059	£31,0	at 10%	£ 3,107
Land Value - as at 1973	£500 per acre		- half acre site	£ 250
				£34,434
			at 5%	£ 1,722
				Net Annual Value

But my proposal of 24 March 1986 proposed Gross Value £2000, Rateable Value £1638, which creates an upper limit constraint on the Court's decision, hence I revise my valuation down to Net Annual Value £1638.

Royal Building
St Andrew's Cross
PLYMOUTH PL1 2DS

Signed M D ELDRED ....
(Chairman)

Date 18 November 1986 ....

APPENDIX 2

SCHEDULE B

THE WIND ENERGY GROUP

Wind Turbine Generator, Slade, Ilfracombe, Devon

Local Valuation Court - Friday, 17th October, 1986

Valuation under Class 1A of the Plant and Machinery (Rating) Order 1960 (as amended) submitted on behalf of the Wind Energy Group (objectors) by Peter Fraser, F.R.I.C.S.

1986 Cost of 70 kw Perkins Diesel Engine Powered Generator.

(Trailer mounted and weatherproofed to include contacter and starter)

Budget Price		£8,514
Notional 1973 cost applying a factor of 0.22		£1,873
Add Notional 1973 capital value of site (based on assumed 1973 capital value of £500 per acre		£250
Effective Capital Value		£2,123
	say	£2,000
Decapitalise at 5% Rateable Value		£100

Matthews & Son,
Chartered Surveyors
91, Gower Street,
London, WC1E 6AB

Royal Building
St Andrew's Cross
PLYMOUTH PL1 2DS

Signed M D ELDRED ....
(Chairman)

Date 18 November 1986 ....

Chapter 5

International energy agency: Recommended practices 4. Measurement of noise emission from wind turbines

S. Ljunggren and A. Gustafsson

FOREWORD

The evaluation of wind turbines must encompass all aspects of a Wind Energy Conversion System (WECS) ranging from: energy production, quality of power, reliability, durability and safety, through to cost effectiveness or economics, noise characteristics, impact on the environment and electromagnetic interference. The development of internationally agreed upon evaluation procedures for each of these areas is needed now to aid in the development of the industry while strengthening confidence and preventing chaos in the market.

It is the purpose of the proposed recommendations for wind turbine testing to address the development of internationally agreed upon test procedures which deal with each of the above noted aspects for characterizing WECS. The IEA expert committee will pursue this effort by periodically holding meetings of experts, to define and refine consensus evaluation procedures in each of the areas:

1. Power Performance
2. Cost of Energy from WECS
3. Fatigue Evaluation
4. Acoustics
5. Electromagnetic Interference
6. Structural Safety
7. Quality of Power
8. Glossary of Terms

This paper addresses the fourth item - Acoustics - and updates the previous publication produced in 1984. The recommendations will be regularly reviewed in the light of the latest knowledge, and areas requiring further investigation will be identified.

SCOPE AND FIELD OF APPLICATION

This document describes the procedures to be used for the measurement and description of the noise emission of wind turbines. The primary goal of the document is to facilitate comparisons of noise measurements made in different countries by different investigators. The secondary goal is to provide an engineering data-base for the development and validation of analytical acoustic prediction techniques.

The document does not address the psycho-acoustic aspects of the acoustic problem, nor does it attempt to define acoustic limits of acceptability for regulatory purposes.

INTRODUCTION

The purpose of this guide is to recommend sound measurement procedures which would enable noise emission of a wind turbine to be characterized. This involves using measurement methods appropriate to noise emission assessment at locations close to the machine, in order to avoid errors due to anomalous sound propagation, but far enough away to allow for finite source size. The procedures described are different in some respects from those that would be adopted for sound propagation in community noise studies. They are intended to facilitate wind turbine noise characterization in a manner suitable for input to noise propagation calculations and for comparisions with the technical specification of a given machine. Standardization of measurement procedures will also facilitate comparisons between different wind turbines.

Several levels of technical capability are allowed for in the guide, depending on the type of wind turbine, but even at the minimum capability level it is expected that good quality measuring equipment will be available or alternatively, assistance can be obtained from a regulatory, research or commercial organization.

RECOMMENDED PRACTICE FOR MEASUREMENTS OF NOISE EMISSION

1. DEFINITIONS

1.1 A-weighted sound pressure level, L_{pA}, in decibels:

The value of the sound pressure level determined using frequency-weighting network A (see IEC Publication 651, reference [1]). The reference sound pressure is 20 μPa.

1.2 Equivalent continuous sound pressure level, $L_{eq,T}$, in decibels:

Value of the sound pressure level of a continuous steady sound that, within a specified time interval T, has the same mean square sound pressure as a sound under consideration whose level varies with time. It is given by the formula

$$L_{eq,T} = 10 \lg \left\{ \frac{1}{t_2 - t_1} \int_{t_1}^{t_2} \frac{p^2(t)}{p_0^2} \, dt \right\} \qquad (1)$$

where

$L_{eq,T}$ is the equivalent continuous sound pressure level, in decibels, determined over a time interval T starting at t_1, and ending at t_2,

p_0 is the reference sound pressure in micropascals (=20 μPa),

p(t) is the instantaneous sound pressure of the noise signal in micropascals.

1.3 Equivalent continuous A-weighted sound pressure level, $L_{Aeq,T}$, in decibels:

Value of the A-weighted sound pressure level of a continuous, steady sound that, within a specified time interval T, has the same mean square sound pressure as a sound under consideration whose level varies with time. It is given by the formula

$$L_{Aeq,T} = 10 \lg \left\{ \frac{1}{t_2 - t_1} \int_{t_1}^{t_2} \frac{p_A^2(t)}{p_0^2} dt \right\} \quad (2)$$

where

$L_{Aaq,T}$ is the equivalent continuos A-weighted sound pressure level, in decibels, determined over a time interval T starting at t_1 and ending at t_2,

p_o is the reference sound pressure (20 μPa);

$p_A(t)$ is the instantaneous A-weighted sound pressure of the sound signal.

Note
In American ANSI-standards, the equivalent continuous A-weighted sound pressure level is called the average sound level with the symbol L_{eq}.

1.4 A-weighted percentiles L_{10}, L_{90} and L_{95} in decibels:

The percentile L_n is defined as the A-weighted sound pressure level that over a specified time is exceeded n per cent of the time. In this document, the percentiles L_{10}, L_{90} and L_{95} are used.

1.5 Reported standard sound level, L_{std}, in decibels:

The value of the equivalent continuous A-weighted sound pressure level corrected to a slant distance of 100 m according to equation (4).

2. INSTRUMENTATION

2.1 Instruments

2.1.1 Equipment for the determination of the equivalent continuous A-weighted sound pressure level.

This equipment shall meet the requirements of a type 1 sound level meter according to IEC Publication 651, reference [1]. The microphone shall have the maximum diameter of 13 mm.

Additional requirements for the sound level meter are that the equipment shall be capable of providing a read-out display or have the facility to be able to obtain the equivalent continuous A-weighted sound pressure level.

If the noise is steady over the period of interest, the measurements may be carried out with equipment that cannot register the true equivalent continuous sound pressure level. In this case the metered "slow" response should be used. The reading should be taken as the average meter deflection. If the meter reading fluctuates over a range greater than 5 dB, the noise is not considered to be steady.

2.1.2 Equipment for the determination of the A-weighted percentiles L_{10}, L_{90} and L_{95}

In addition to the requirements given in clause 2.1.1, the equipment shall be capable of providing a read-out of, or otherwise have the facility to obtain the A-weighted percentiles L_{10}, L_{90} and L_{95} with the time weighting "F" (fast) according to IEC 651.

2.1.3 Equipment for the determination of third-octave band spectra

In addition to the requirements given in clause 2.1.1, the equipment shall have a substantially constant frequency response over at least the frequency range 45 Hz to 5600 Hz. The filters shall meet the requirements of IEC Publication 225, reference [2].

The equipment shall be capable of providing a read-out of the equivalent continuous sound pressure level in third-octave bands with centre frequencies from 50 Hz up to 5000 Hz.

2.1.4 Equipment for the determination of narrow band spectra

In addition to the requirements given in clause 2.1.1, the equipment shall have a substantially constant frequency response from 45 Hz up to 5600 Hz.

For analyzers with constant absolute bandwidth, the noise bandwidth (the effective bandwidth) shall be not wider than approximately 7 Hz for frequencies below 2000 Hz and not wider than approximately 70 Hz for frequencies above 2000 Hz. For analyzers with constant relative bandwidth (constant percentage bandwidth), it shall be possible to choose a noise bandwidth in the region of 3% (1/24 octave).

The averaging time shall be at least 2 minutes and the average shall be taken of at least 100 spectra.

2.2 Microphone with reflecting surface and windscreen

The microphone shall be mounted on a hard board with the diaphragm of the microphone in the vertical plane and with the axis of the microphone pointing towards the wind turbine, see Figure 1. The board shall have an area of at least 1.5 m x 1.8 m and be made from a material that is acoustically hard, e.g., a piece of plywood or hard chip-board with a thickness of at least 16 mm. The board shall be placed on the ground with its long sides pointing towards the wind turbine.

The windscreen to be used together with the ground-mounted microphone consists of a primary and, where necessary, a secondary windscreen. The primary windscreen consists of one half of an open cell foam wind screen with a diameter of approximately 90 mm, which is centred around the diaphragm of the microphone, see Figure 1. The secondary windscreen should be used when necessary to obtain an adequate signal-to-noise ratio in high winds. It consists of a wire hemispherical frame, at least 450 mm in diameter, which is covered with a 25 mm layer of open cell foam with a porocity of 4 to 8 pores per 10 mm. This secondary hemispherical windscreen shall be placed symmetrically over the smaller, primary windscreen.

2.3 Recording of data

When data are stored on tape as an essential part of the measuring procedure, any additional errors caused by the process of storing and replay shall be taken into account when presenting the result of the measurement.

2.4 Calibration

The complete measurement chain must be calibrated at least at one frequency before and after the measurements. An acoustic calibrator with an accuracy of ±0.5 dB shall be used.

3. MEASUREMENTS

3.1 Measuring positions

Five microphone positions are to be used. Four positions shall be laid out in a pattern round the wind turbine with the fifth position further downwind as indicated in Figure 2. The distance from the wind turbine to each microphone position shall be that indicated in Figure 2 with a tolerance of ±20%. The reference distance R_0 is given by (see also Figure 3):

$$R_0 = H + D/2 \tag{3}$$

where, for a horizontal axis wind turbine,

H is the distance from ground to centerline of the rotor shaft,

D is the diameter of the rotor.

For a wind turbine with a vertical axis

H is the highest point of the structure,

D is the equatorial diameter.

The microphone shall normally be mounted on a large board which is positioned on the ground, see clause 2.2. When the propagation path to the wind turbine is obstructed or when it is impractical to use a ground microphone position, e.g. for off-shore wind turbines, a microphone without a board and at the height of 5.0 m may be used.

The measurement position shall be chosen so that the influence of reflecting structures (e.g. buildings) is minimized.

3.2 Acoustic measurements

3.2.1 Equivalent continuous A-weighted sound pressure level

The equivalent continuous A-weighted sound pressure level shall be determined at the five measuring positions and during the following conditions:

A. Wind turbine in operation
   A1. Wind speed as close to cut-in as possible,
   A2. Wind speed as high as possible.

B. Wind turbine parked
   B1. Wind speed as close as possible to the wind speed during the A1 measurement.

The difference in wind speed during the conditions given by A1 and A2 shall be at least 3 m/s.

Each measurement is recommended to be at least 2 minutes in duration where practicable and during periods of steady wind. Remarks on subjective impression of noise (audible discrete tones, impulsive character, spectral content, temporal characteristics, etc.) should be noted.

Measurements shall be carried out simultaneously at the reference position and at least at one of the other points.

3.2.2 Measurements at the reference position

In addition to the measurement of the equivalent continuous A-weighted sound pressure level, the following quantities shall be determined at the reference position:

a) the equivalent continuous sound pressure level in third-octave bands with centre frequencies from 50 Hz up to 5000 Hz,

b) the A-weighted percentiles L_{10}, L_{90} and L_{95},

c) in the case of audible pure tones, a narrow band spectrum covering the appropriate part of the frequency spectrum.

It is recommended that the measurement time period in case a) is at least 15 sec for each frequency band and in case b) is at least 2 minutes.

These measurements shall be carried out a wind speed as near cut-in + 2 m/s (measured at hub height) as possible.

3.3 Non-acoustic measurements

During the measurements specified in clause 3.2 the following wind turbine related parameters shall be continuously recorded:

- wind speed and direction at hub height. The wind speed should preferrably be obtained from measurements of the power produced and the measured power versus wind-speed curve. In the second place it can be measured with an anemometer at hub height and in the third place at 10 m height. The anemometer and wind direction transducer must be located upwind the wind turbine and not closer than 1.5 rotor diameters for horizontal-axis machines and 2 rotor diameters for vertical-axis machines and not farther than 6 rotor diameters.

During the course of the test, the anemometer must never be in the wake of any portion of any wind turbine rotor or structure (Ref [3]).

- power output,
- rotor speed,

and optionally also

- blade pitch angle,
- yaw angle.

Also to be recorded every 1/2 hour are:

- relative humidity,
- temperature,
- barometric pressure,
- turbulence (qualitative assessment). If equipment to measure the turbulence is not available, the following parameters should be reported: time of day, cloud cover and terrain type. The turbulence can then be estimated approximately from these parameters combined with windspeed,
- the possible presence of a temperature inversion in the atmosphere.

4. DERIVED RESULTS

4.1 Reported standard sound levels

As specified in clause 3.1, the measuring positions may differ from R_0 and $2R_0$ respectively by ±20 %. There are also two alternative microphone heights of 0 m and 5 m. In order to obtain comparable results, the equivalent continuous A-weighted sound pressure levels shall be corrected to a reference distance and, when appropriate, to a measurement point on the ground according to the following equation:

$$L_{std} = L_{Aeq,T} + 20lg(R_i/R_N) - K, \qquad (4)$$

where

L_{std} is the reported standard sound level, in decibels,

$L_{Aeq,T}$ is the measured equivalent continuous A-weighted sound pressure level, in decibels,

R_i is, in the case of a horizontal axis wind turbine, the distance from the measurement position to the hub of the yaw axis. In the case of a vertical axis wind turbine, R_i is the distance from the measurement position to a point at the axis of rotation midway between the upper and lower blade attachments, see also Figure 3,

R_N is the reference slant distance, =100 m,

K is

0 dB with microphone on the ground according to clause 2.2,

3 dB when the alternative microphone height of 5 m has been used.

If a single value of the 5 values obtained according to equation (4) is reported, this value shall be the maximum value.

5. ADDITIONAL INFORMATION TO BE RECORDED

5.1 Characterization of the wind turbine

The geometric configuration of the wind turbine and its operating conditions must be completely specified. The wind turbine configuration should be described in detail and should include such information as rated capacity, principal dimensions, hub height, tower type, rotor dimensions and geometry, number of blades, upwind or downwind, overspeed controls, fixed or variable pitch blade angle, teetering rotor, yaw configuration, generator type, geartooth frequency, etc.

5.2 Acoustic environment

The following information on the acoustic environment at and near the site of the wind turbine and the measuring positions shall be recorded:

a) type of topographic terrain (hilly, flat, cliffs, mountains, etc, for nearest 1-2 km). Photographs should be included,

b) type of ground (grass, sand, etc.),

c) nearby reflecting structures such as building structures and sound sources such as trees, bushes, water surfaces, etc.,

d) other nearby sound sources such as other wind turbines, highways, industrial complexes, airports etc., which may affect the background level.

5.3 Instrumentation

- The equipment used for the measurements, including type, serial number and name of manufacturer,

- frequency response of instrumentation system,

- bandwith of narrow band frequency analyzer.

5.4 Acoustic data

- The locations and orientation of the microphone at each measurement position,

- the corrections in decibels, if any, applied in each frequency band for the frequency response of the microphone, frequency response of the filters in the pass band, background noise, microphone height, etc,

- the reference distance R_0.

6. INFORMATION TO BE REPORTED

In addition to the data specified in clause 3 - 5, the following information should appear in the test report:

a) Identification of wind turbine

The test report should contain a complete description of the wind turbine which was tested. Information to be included should contain a description (and if appropriate, a diagram) of the appearance of the wind turbine, its conventional operating characteristics, and the performance characteristics of the machines which may relate to noise. The power curve should be included.

b) Wind turbine operating procedure

A description of how the wind turbine is operated under test is an essential part of the report. If the device is operated in any manner other than its conventional operating mode, this operating procedure should be completely described.

c) Acoustic data

The reported standard sound levels from the five measurement positions rounded to nearest whole decibel. The third-octave band spectra and the percentiles obtained at the reference point.

d) Measurement uncertainty

The section on measurement uncertainty should give some indication of the degree of confidence which can be placed in the measurement results. This could be expressed in terms of standard error, confidence limits, or some other appropriate statistical factor.

e) Time and date

Time and date when the measurements were performed.

7. ACKNOWLEDGEMENTS

The present edition of this document has been developed through a series of meetings with participants from different countries participating in the IEA R&D-agreement. The following persons have participated with valuable contributions at these meetings

B. Andersen and J Jakobsen, Denmark
J. van der Toorn and L. van Schie, The Netherlands,
A. Robson, R. Dunlop and A. R. Henderson, United Kingdom,
S. Meijer, Sweden.

The appendices have been written by S. Meijer and A. Gustafsson (No 1), B. Andersen (No 2), J. Jakobsen (No 3), S. Ljunggren (No 4) and A. Gustafsson and S. Meijer (No 5).

During the work leading to this revision, an informal contact has been maintained with the acoustics subcommittee of the American Wind Energy Association (AWEA). This contact has resulted in a certain co-ordination of the present IEA document with a future AWEA/ANSI standard. It has not been possible to obtain perfect co-ordination, but it is thought that the similarities are fundamental enough to render comparisons possible between measurement results obtained using the two procedures.

8. REFERENCES

[1] IEC Publication 651. Sound level meters.

[2] IEC Publication 225. Octave, half-octave and third-octave band filters intended for the analysis of sound and vibrations.

[3] Expert Group Study on Recommended Practices for Wind Turbine Testing & Evaluation: 1. Power Performance Testing; Edited by Sten Frandsen; B. Maribo Pedersen, submitted to Executive Committee (the International Energy Agency Programme for Research & Development on Wind Energy Conversion Systems). Revised Edition 1988.

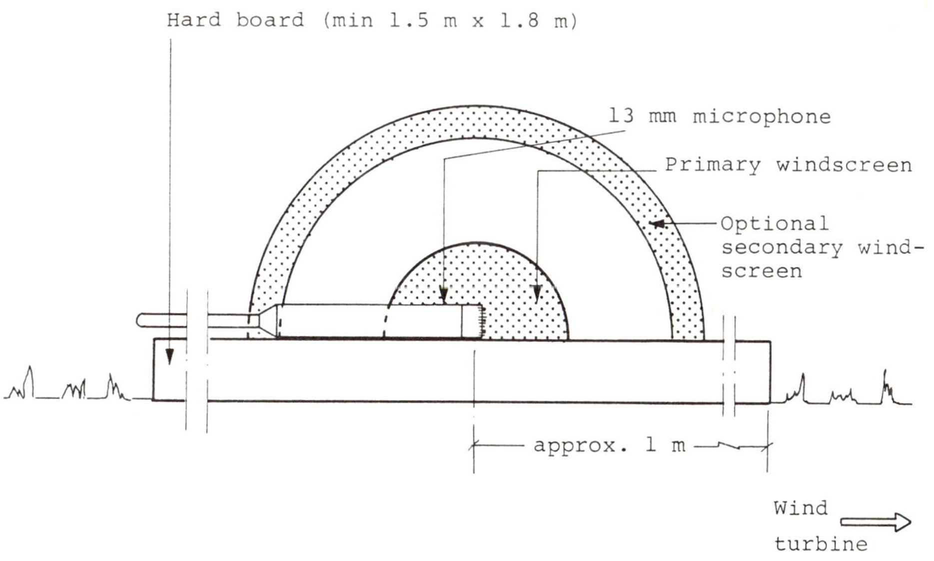

Figure 1. Mounting of microphone. (Not to scale)

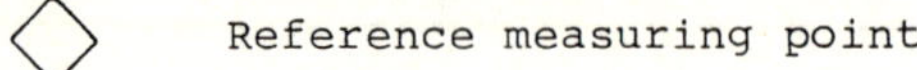

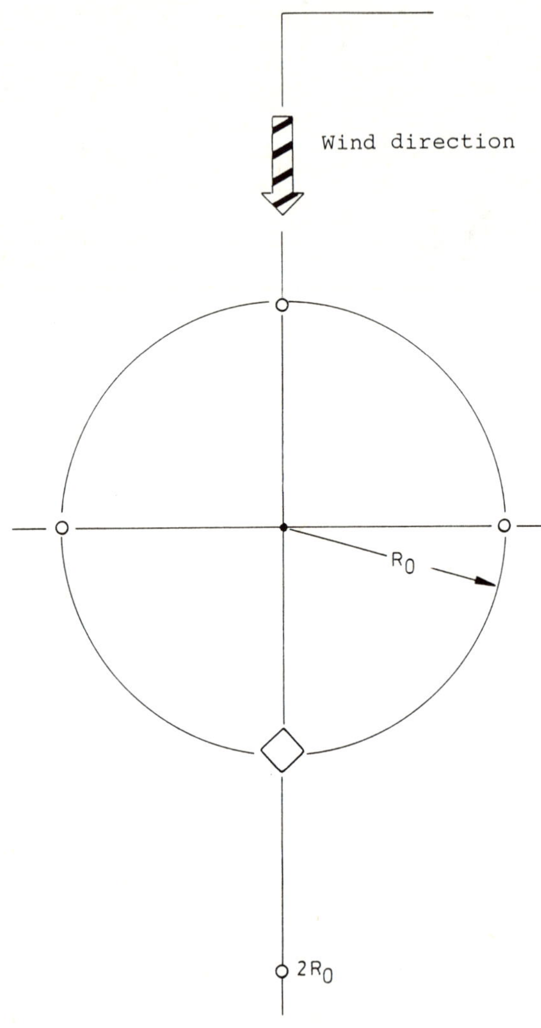

Figure 2. Recommended pattern for measuring points

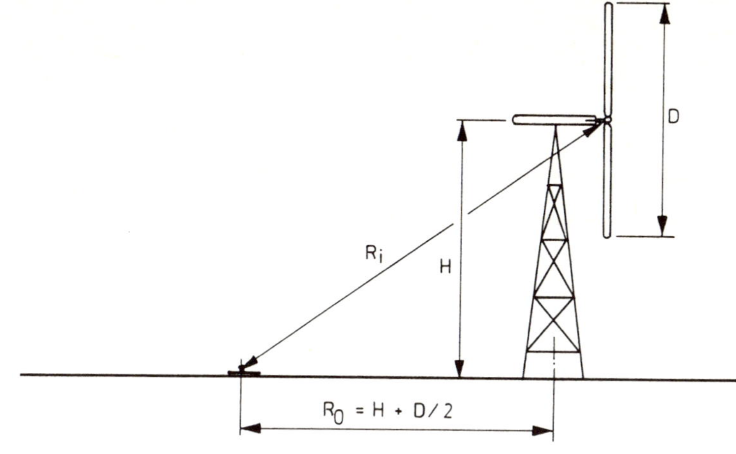

a)

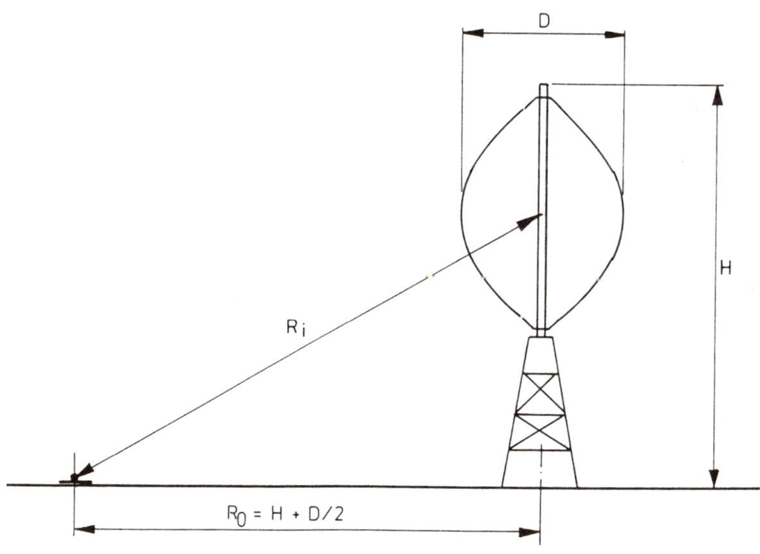

b)

Figure 3. Illustration of the definitions of R_0 and R_i for a) horizontal-axis turbine
b) vertical-axis turbine

APPENDIX 1.

SUPPLEMENTARY MEASUREMENTS RECOMMENDED FOR TURBINES WITH PERIODIC UNSTEADY BLADE LOADS

An impulsive, thumping, sound may be heard from wind turbines with periodic, unsteady loads on the rotor blades. These unsteady loads can be due to a number of causes e.g. wakes from support tower structural members for downwind and vertical axis turbines. The sound pressure level spectrum contains distinct spectral peaks at harmonics of the blade passage frequency.

To assess, in a simple way, whether noise from the wind turbine exhibits impulsive behaviour a method described in [1] can be used. In this method the pressure signal is filtered in a 31.5 octave band filter and converted to a d. c. voltage by a precision sound level meter operating in the "fast" response mode. The signal is displayed on a stripchart recorder or on a spectrum analyzer with a time domain display. It is often easier to discern the impulsive behaviour of the noise from the filtered signal than from the unfiltered one.

Narrow band spectra for the frequency range f_L to 100 Hz, where f_L is in the interval 10 Hz - 20 Hz, and with a bandwidth smaller than half the blade passage frequency should be determined and reported together with examples of time series traces of the pressure signals.

The characteristics of the impulsive sound can be influenced by the local atmospheric conditions and as a consequence of this the sound measurements should if possible be made after sunset.

Wind conditions, measurement points and recording of non-acoustic data should be as specified in the main text.

<u>Reference</u>

[1] Kelley N.D. A Suggested Method for Field Assessment of Impulsive Noise Characteristics of Wind Turbines prepared for the Acoustics Standards Committee, American Wind Energy Association May 11, 1987.

APPENDIX 2.
THE EFFECT OF THE MICROPHONE MOUNTING METHOD ON THE MEASURED SOUND PRESSURE LEVEL

The effect of the hard board was tested in a small experiment, where noise from a reference sound power source (B&K 4205) was measured by two microphones on the board (one flush mounted, the other "parallel"-mounted). The hard board was placed upon a lawn. Immediately after these measurements the microphone systems were placed in a large anechoic chamber (without the board, noise incidence in the plane of the microphone membranes-angle of incidence = 90°). Figure 1 a, b and c shows the difference between the 1/3-octave band levels measured upon the hard board and in free field with the angle of incidence as parameter.

The deviations from the expected +6 dB can be explained by the impedance jump approximately 1 m in front of the microphones.

The effect of simply laying the microphone upon the hard board ("parallel"-mounted), instead of actually flush-mounting it, was investigated using a B&K 4134 pressure microphone flush mounted and a B&K 4165 free-field microphone "parallel"-mounted in the above mentioned experiment. In the case of the "parallel"-mounted microphone, the microphone was mounted with an azimuth angle towards the source of either 90° or 0°.

Figure 2 shows roughly the two mounting-methods, while Figure 3 shows the error using the "parallel"-mounting method. The figure shows the difference between the 1/3-octave band levels measured with the flush-mounted and the "parallel"-mounted microphone. The full line corresponds to noise incidence in the membrane plane of the "parallel"-mounted microphone (azimuth angle 90°).

For an angle of incidence in the range 30° - 60° (in the vertical plane) the error at 8 kHz will be -1 dB to -7 dB using a "parallel"-mounted microphone pointing towards the sound source.

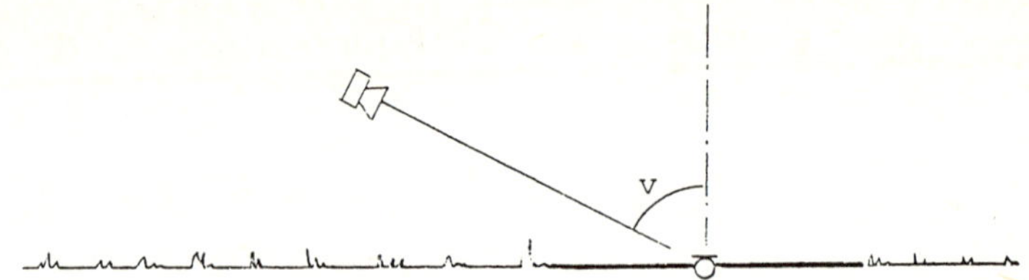

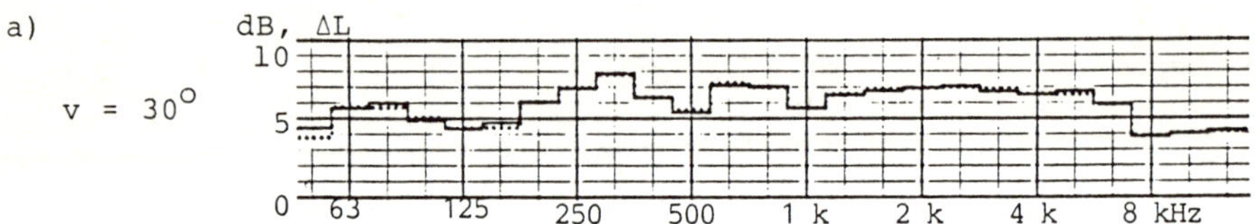

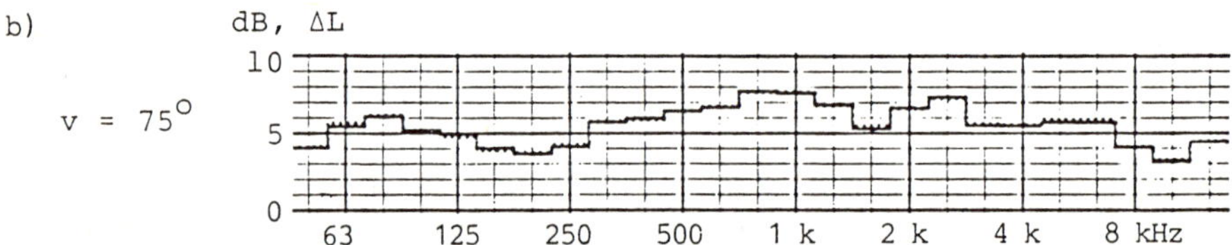

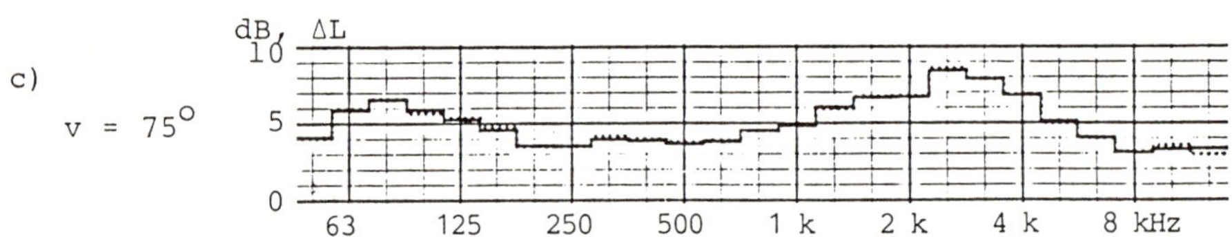

Figure 1: Difference (ΔL) between 1/3 octave band spectra of noise from reference sound power source measured with a flush-mounted microphone upon the hard board on a lawn and the same microphone in free field (anechoic chamber). v indicates the angle of sound incidence and ... a repetition of the measurement.

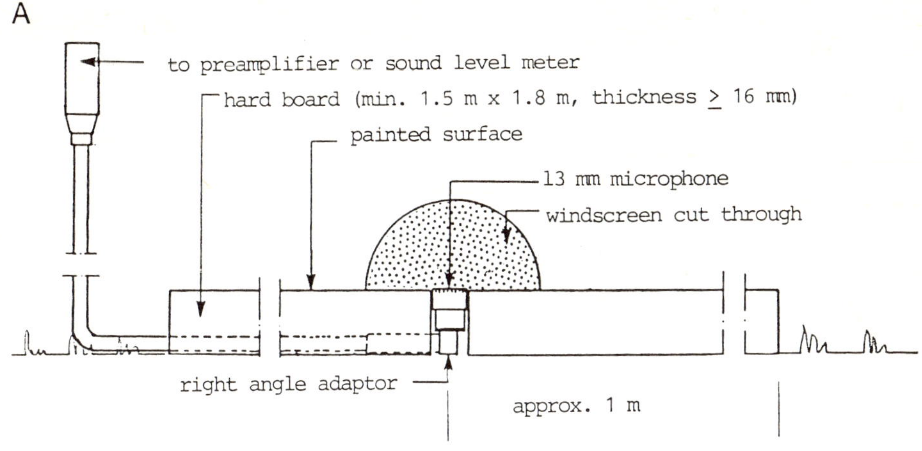

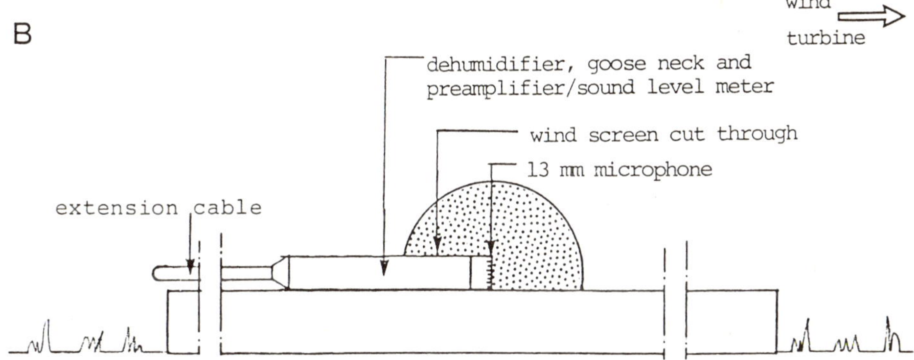

Figure 2: Examples of microphone systems using a hard chip- or plywood-board.
Set-up A: Flush-mounted microphone
Set-up B: "parallel"-mounted microphone

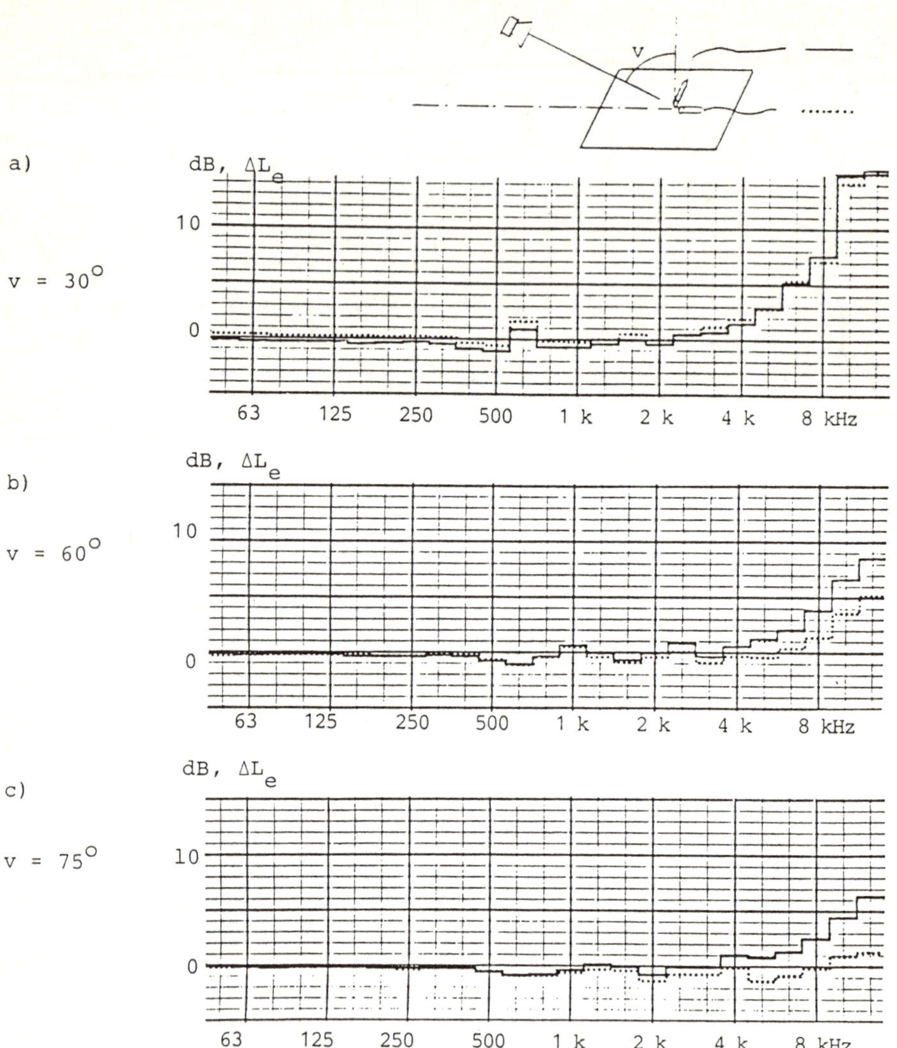

Figure 3: Under-estimation using microphone set-up B instead of A. Difference (ΔL_e) between 1/3-octave band spectra measured with flush-mounted and "parallel"-mounted microphone. v indicates the angle of sound incidence. —— in the membrane plane and ... microphone pointing towards the sound source.

APPENDIX 3.
PREDICTION OF NOISE LEVELS AROUND A WIND TURBINE

This appendix describes a method to predict the noise from a wind turbine in its surroundings.

Often it is very dificult to measure the noise from a wind turbine at long distance due to background noise - mostly from the wind acting upon trees and bushes - and to wind-generated microphone noise. In such cases and in a planning situation where the siting of a wind turbine is investigated, it is advantageous to predict the noise.

1. From the sound pressure levels close to the wind turbine the sound power level can be calculated. When the measurements are carried out at longer distances from the tower base than one hub height (preferably between one and two hub heights), the wind turbine can be regarded as a point source.

   The measured sound pressure levels are corrected for the ground effect. In this way the free-field sound pressure levels are obtained. The sound power level is:

   $$L_W = L_{p,free} + 10lg(4\pi R_i{}^2)$$

   R_i is the distance between measurement point and hub,

   $L_{p,free}$ is the average of free-field sound pressure levels at this distance,

   In many cases it is relevant to predict the noise from the wind turbine in the downwind direction only. Such cases exist e.g. when noise limits apply to the maximum noise rather than to the long-term average. In such cases the "downwind" sound power level is calculated on the basis of one or a few sound pressure levels measured in the downwind sector.

The correction for the ground effect can be made when the ground is totally reflecting (concrete, asphalt, or the like). When the microphone is situated at a finite height above the reflecting ground, the free-field sound pressure level is estimated as:

$$L_{p,free} = L_p - 3 \text{ dB}$$

When the microphone is very close to a reflecting surface placed upon the ground (preferably flush-mounted in this surface), the free-field sound pressure level is found with reasonable accuracy as:

$$L_{p,free} = L_p - 6 \text{ dB}$$

Estimates of the free-field sound pressure level based on sound pressure levels measured above porous (acoustically soft) ground is subject to a considerable uncertainty and is not recommended.

2. On the basis of the sound power level or the "downwind" sound power level, the sound pressure level in the downwind direction at the distance R from the hub is found as:

$$L_p(R) = L_W - 10\lg(2\pi R^2) - \Delta L_a$$

ΔL_a is the correction for air absorption. It can be calculated as $\Delta L_a = R x \alpha_a$, where α_a is shown for standard atmospheric conditions in Table 1 below.

1/1 octave centre frequency, Hz	63 & 125	250	500	1000	2000	4000	8000
α_a, dB/m	0.000	0.001	0.002	0.004	0.007	0.017	0.056

Table 1 Coefficients for sound absorption in the air. Valid for T = 15°C, RH = 70%. Reference made to e.g. ANSI S 1.26 - 1978 "Method for calculation with absorption of sound by the atmosphere".

A limited number of measurements around small and medium-sized wind turbines (P: 50-75 kW, hub height: 22-30 m, wind speeds: 6-10 m/s at 10 m height) have shown that noise levels predicted in this way are in reasonable agreement with measured noise levels (deviations of A-weighted sound pressure levels were generally within ±2 dB). It was also found that serious differences occurred when the prediction was instead based on a more detailed prediction method, taking the effect of the porous ground into consideration e.g. Kragh et al. "Environmental noise from industrial plants. General prediction method". Danish Acoustical Institute, Report No. 32, Lyngby 1982 .

APPENDIX 4
MEASUREMENT OF BACKGROUND SOUND

The background sound at a wind turbine site is usually generated in several different ways. Typical examples are aerodynamic sound from the interaction of the wind with buildings and trees, rustle of leaves, sound from birds and other animals, sound from human activities, noise from factories, cars and trucks, trains, aircraft etc. Of these different types, the sound generated by the wind is of particular interest as it has a frequency spectrum that is similar to that of the broadband noise generated aerodynamically by a wind turbine. This implies that the noise from a wind turbine can be masked by natural wind sound. Thus, a relevant assessment of the wind sound is of particular interest.

As the background sound is important for the masking of the noise from a wind turbine, it is of utmost importance that the background sound is measured at relevant points, i.e., the same points where the noise from the wind turbine is deemed to be important. In many cases this will lead to measurement points near existing or future houses. It should be observed, however, that the existance of a new house at a site way well change the background sound situation.

It is recommended that the same measurement technique is used as that described in Annex 2 and Chapter 2 of the main text in order to minimize the wind-induced noise from the microphone. The measurements should be carried out at the two wind speeds described in clause 3.2.1 of the main text. The wind speed should be measured at a point or points relevant for the generation of the background sound. A measurement height of 10 m is recommended.

APPENDIX 5
A COMMENT ABOUT TURBINES WITH A VARIABLE ROTATIONAL SPEED

The aerodynamically generated noise is strongly dependent on the rotational speed of a wind turbine. The sound level in dB(A) generally varies as 50lg(RPM) where RPM is the number of revolutions per minute.

Measurements should be made at a wind speed close to where the turbine reaches its maximum rotational speed.

For turbines with fast response of the rotational speed to variations of the wind speed through the rotor care must be taken if attempts are made to correlate the noise with a wind speed not measured close to the rotor.

At each measurement point, the same acoustical quantities should be obtained as at the reference point when the noise emission from the wind turbine is determined, see clause 3.2.2. As there still are many unclear points in connection with the assessment, and especially the masking effect, of the background sound it is recommended that tape recordings be made and saved for future evaluations.

It is strongly recommended that photographs of the surroundings of the measurement points are included in the report.

Chapter 6

Propagation of noise from wind turbine generators

J. N. Pinder, M. A. Price and M. G. Smith

SYNOPSIS

Most prediction models for outdoor sound propagation have been designed and validated for application at very low wind speeds, below the onset of WTG operation. A propagation model is therefore under development applicable to WTG operations in order to assist the development of guidelines for the assessment of noise received in communities near WTG sites. Ray tracing theory has been applied, assuming an acoustically "soft" ground plane and flat ground. The algorithm adopted aims to avoid the requirement for heavy use of computer resources.

1. INTRODUCTION

Outdoor sound propagation has been studied, theoretically and experimentally, for many years. Such studies have usually been tailored to answer particular questions and to apply to a limited set of conditions. Sound propagation in windy conditions is imperfectly understood since, until the advent of wind energy programs, there has been little need to study this phenomenon. Most practical noise prediction schemes are restricted to low windspeed situations, below about 6 m/s, acknowledging the difficulties in making sound measurements in windy conditions and the variability of perceived noise in wind. It is evident that low windspeed prediction schemes are not valid for wind turbine noise and thus a different approach is required which will take better account of the effects of wind. Much interest has been shown in the effects of temperature gradients on sound propagation, but this is thought to be of little importance for wind turbine noise, since strong temperature gradients are understood not to develop in windy conditions.

2. ANALYTICAL MODELLING OF NOISE PROPAGATION FROM WIND TURBINES

There are several theoretical methods of modelling noise propagation in a realistic non-uniform and/or moving atmosphere. These include wave methods, finite element modelling and ray theory methods. The first two methods, whilst potentially providing accurate answers (particularly in the near field), become overwhelmingly intensive of computer time when used to study propagation over large distances. Ray theory, by making simplifying assumptions about the sound field and the length scale of changes in the atmosphere, provides a much more manageable form of solution.

A ray is a line which is perpendicular to the wavefront of the sound at all points along the ray path. It is possible to formulate a so-called dispersion relationship which governs the ray paths. Using this relationship paths of individual rays can be found without determining the wavefronts throughout the sound field, as would be required by wave theory.

Figure 1a shows how, with no wind or temperature gradients, rays propagate in straight lines so that there is one direct and one indirect (reflected) ray path between source and observer. With wind and temperature gradients, however, there may be more than two ray paths for a downwind observer (figure 1b) or fewer than two rays upwind (figure 1c). The sound field in a non-uniform atmosphere may also be affected by focussing effects caused by the refraction of the wavefronts. A model is under development which aims to predict the sound field around a source in windy conditions using ray theory.

The method used in the model employs 'ray-tubes' to predict the sound intensity at points along the ray. The cross-sectional area of the ray-tube is proportional to sound intensity and can increase or decrease depending on the refractive conditions (see Figure 2). This method is in contrast to some other workers in the field who trace many rays and find the noise intensity from the spread of neighbouring rays. The model assumes a continually varying atmosphere with a wind gradient of the form:-

$$u = u_o \left[\frac{Z}{Z_o} \right]^{\xi}$$

where u = windspeed at height Z

u_o = windspeed at height Z_o

ξ depends on the stability of the atmosphere

$0.14 < \xi < 0.77$

0.14 for superadiabatic lapse condition

0.77 for a highly stable inversion.

A low value is indicated for high wind conditions.

Rays are traced by performing a numerical integration following the method of Candel [1]. The governing differential equations of the ray paths given by Candel include the effects of wind speed, wind gradients, temperature gradients and reflection from solid boundaries such as the ground. The temperature gradient calculations have not been included in the model under development. Flat ground is assumed in the present model.

Another development in the present application is the way in which rays passing through both the source and observer ('eigenrays') are found. Some methods have used a large number of rays emanating from the source and selected those that pass near the observer. Information gained from the ray-tube calculation, however, may be used in an iterative process to find the actual rays which pass between the source and observer and this is the method chosen. This technique considerably reduces the required computing power; the whole program can be run on an IBM PC-type computer.

There are two parts to the calculation of the sound field. The first is to find all rays propagating between source and observer. The second part is to calculate and sum the sound intensity reaching the observer along each ray, taking into account atmospheric absorption and absorption on reflection at the ground.

3. PREDICTED BEHAVIOUR UPWIND

At short distances and relatively high windspeeds, the main effect of the wind is the formation of a shadow zone upwind of the source. In windy conditions, the model identifies a zone upwind of the source and close to it where two ray paths exist. At a further distance from the source, there is a zone where only a direct ray reaches the observer and beyond that lies the shadow zone which is not penetrated by any ray paths. In practice, a shadow zone does not represent zero sound propagating from the source since other effects such as diffraction and the scattering of sound due to turbulence degrade the shadow zone. Shepherd and Hubbard [4] suggest that attenuation in a shadow zone increases linearly with distance into the shadow zone up to twice the source-to-shadow zone distance, where it reaches a maximum value of between 6 and 30 dB depending on frequency (see Figure 3).

4. PREDICTED BEHAVIOUR DOWNWIND

Figure 1b illustrates the effect of a wind gradient on the ray paths of the sound propagating downwind from the source, with an increasing number of ray paths occurring at longer distances. Close to the source there is a zone with two ray paths, similar to the no-wind situation. The increased number of rays causes an enhancement of sound received compared with the no-wind case, ie. reduced attenuation. Hawkins [2] describes the 'trapping' of sound rays close to the ground as a sound channel which has an attenuation rate of 3 decibels per doubling of distance (dB/dd) rather than 6 dB/dd due to spherical spreading expected in still conditions. The ray theory used here exhibits a pattern of jumps in the attenuation vs. distance graph as more rays appear. Hawkins also predicts this behaviour and illustrates that the ray theory matches well to measurements made downwind of a large WTG at very low frequencies (8-11 Hz). This is interesting since ray tracing is essentially a high frequency approximation to wave theory. At such low frequencies the effects of atmospheric and ground absorption are small.

5. MEASURED SOUND PROPAGATION FROM WTG'S

In the USA, there has been much interest in the low frequency content radiated by WTG's with the rotor downwind of the support tower. Measurements [4] have shown that low frequency WTG noise is reduced by 3 dB/dd downwind, whilst upwind of the source the low frequency noise attenuates at a rate of 6 dB/dd. The ray theory prediction of an upwind shadow zone is not apparent at low frequencies and the attenuation is essentially as if it were due to spherical spreading. Hubbard and Shepherd [4] found that in measurements made up to 1 km downwind of a large wind turbine, sound levels at the higher frequencies, above 63 Hz, decay at a rate equivalent to the effects of spherical spreading and atmospheric absorption. This increase in attenuation over the 3 dB/dd predicted by the sound channel could be due to the effect of ground absorption which is significant at higher frequencies. Hawkins explains a similar result by assuming a higher effective source height for the high frequency part of the noise.

Much of the literature concerning wind turbine generator noise in Europe concentrates on the noise emission of the machines and thus on measurements made close to them. Results reported [3] of noise measurements from two WTG prototypes at distances of up to 400m (about 5 diameters) show little difference between levels upwind and downwind and between results in wind and predictions for the no-wind case for a similar source strength. These results may have arisen because the wind conditions and tower height were such that the shadow zone and downwind zones of enhancement were at greater distances than the measurement.

In hilly terrain, the use of ray tracing may illustrate areas of so-called 'focussing' of sound. It is likely that such effects are due to reflections from various surfaces such as valley sides, which combined with the refraction effects of wind gradients cause an enhancement of sound pressure in certain locations.

In conclusion, the evidence reported thus far suggests that the actual focussing of sound within ray tubes is of minimal effect in the propagation of sound in windy conditions. The present study predicts that enhancement relative to spherical spreading may occur due to the existence of more than two ray paths by which sound may reach the observer, caused by the effect of terrain or refraction of sound downwind. The main upwind effect is the formation of a shadow zone due to refraction of sound upwards away from the ground.

REFERENCES

[1] CANDEL, S. (1977). Numerical solution of conservation equations arising in linear wave theory. J. Fluid Mech, 83.

[2] HAWKINS, J.A. (1987). Application of ray theory to propagation of low frequency noise from wind turbines. NASA CR-178367, Applied Research Laboratories, The University of Texas at Austin.

[3] LJUNGGREN, S. (1984). A preliminary assessment of environmental noise from large WECS, based on experiences from Swedish prototypes. FFA report TN 1984-48.

[4] SHEPHERD, K.P. and HUBBARD, H.H. (1985). Sound propagation studies for a large horizontal axis wind turbine. NASA CR-172564.

ACKNOWLEDGEMENT

The Wolfson Unit for Noise and Vibration Control, ISVR, is being supported in this work by the Department of Energy, Great Britain, through the Energy Technology Support Unit, Harwell Laboratory.

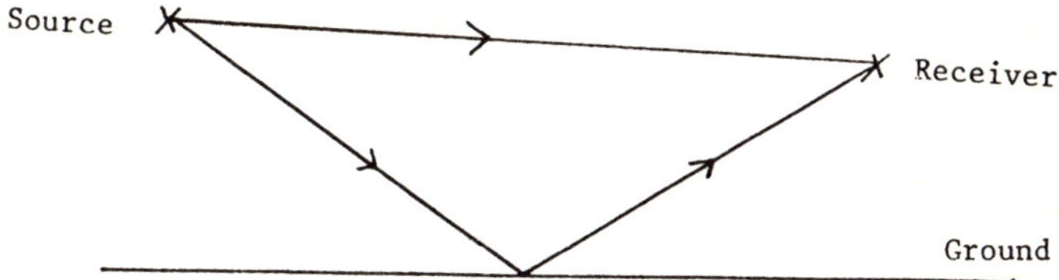

Figure 1a. Propagation in the Absence of Wind:
2 ray paths between source and receiver.

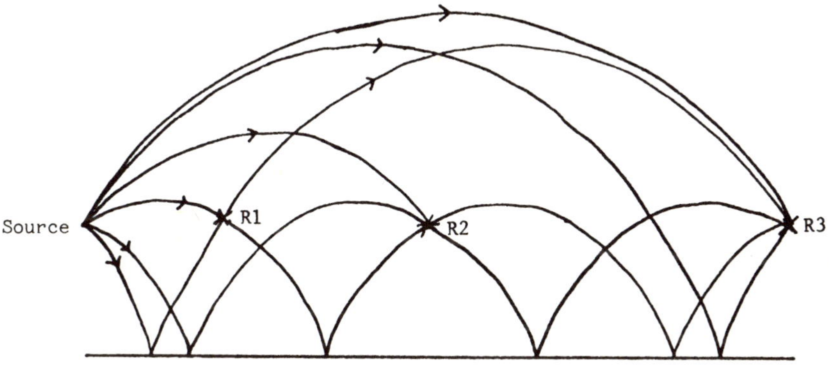

Figure 1b. Downwind Propagation.
Receivers: R1 2 ray paths, R2 3 ray paths,
R3 5 ray paths.

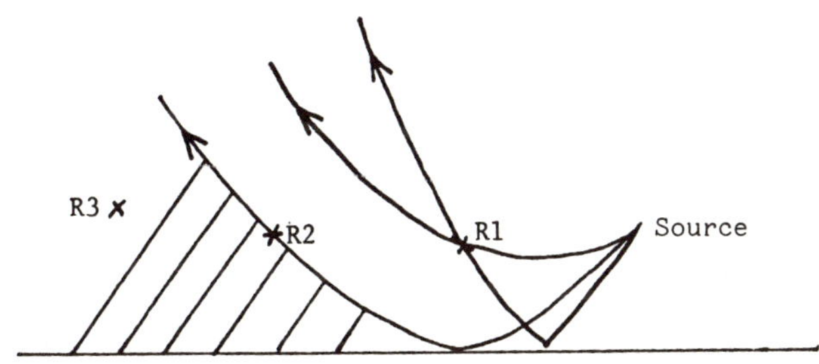

Figure 1c. Upwind Propagation.
R1 2 ray paths, R2 1 ray path, R3 shadow zone,
no ray paths.

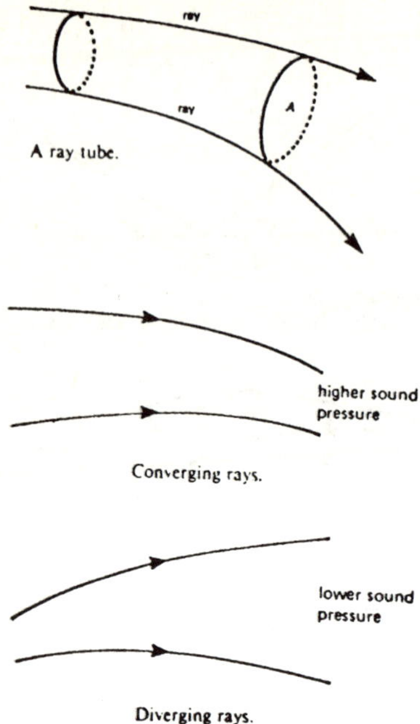

Figure 2. Ray Tubes

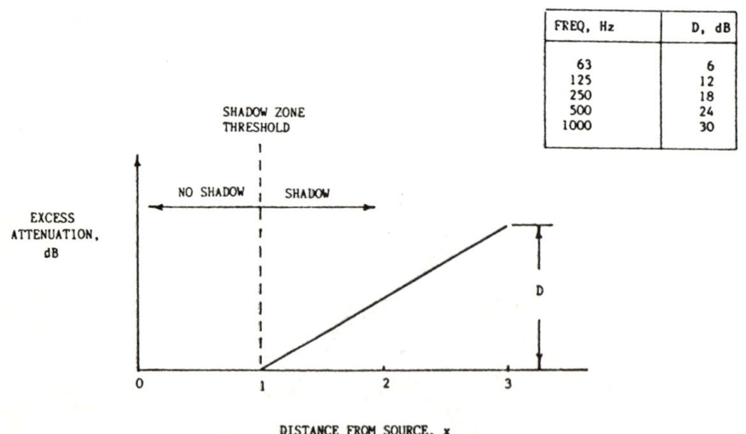

FREQ, Hz	D, dB
63	6
125	12
250	18
500	24
1000	30

Figure 3. EXCESS SHADOW ZONE ATTENUATION AS A FUNCTION OF DISTANCE FROM A POINT SOURCE.
(from [4])

Chapter 7

A Swedish solution to the noise problem: wind turbines with variable RPM

S. Ljunggren and S. Engström

SYNOPSIS The use of a turbine rotor speed which is proportional to the wind speed is discussed with regard to the masking of the turbine noise by natural wind sound.

1. INTRODUCTION

The Swedish government recently decided upon a reduction of the number of operating nuclear reactors in Sweden. The decision gives cause for a study of a large-scale installation of wind-power capacity. Thus, a hypothetic case with a total capacity of 30 TWh/year has been studied. Two thirds of the output is presumed to originate from offshore based turbines and the rest from landbased systems, mainly in the coastal regions. It can readily be shown that if turbines in the Megawatt class are used, the required energy output in conjunction with the scarcity of suitable sites imply that the distance between the landbased turbines and adjacent populated areas cannot be more than approximately 300 - 500 m. This would lead to sound levels of about 50 - 60 dB(A) at the living areas, if no measures are taken against the noise. On the other hand, the Swedish guide-lines for noise from industrial plants set a limit of 35 - 40 dB(A). These guide-lines are not applicable to wind turbines directly, but they still show clearly that the noise from the turbines is far too high in this case, maybe with so much 15 - 20 dB.

A solution to the problem could be to take advantage of the masking effect from the natural wind sound from trees, shrubs, buildings, etc. For this reason, it is presupposed that there are high trees near the relevant houses, that the wind turbines are of the upwind, horizontal axis type without audible mechanical noise, and that the speed of the rotor is proportional to the wind speed below a certain rated wind speed and constant above this speed. This leads to the noise situation which is illustrated in Figure 1.

The concept is based on some measurements at the two Swedish prototypes of the spectrum shapes, influence of rotor speed, amplitude modulation, and temporal characteristics.

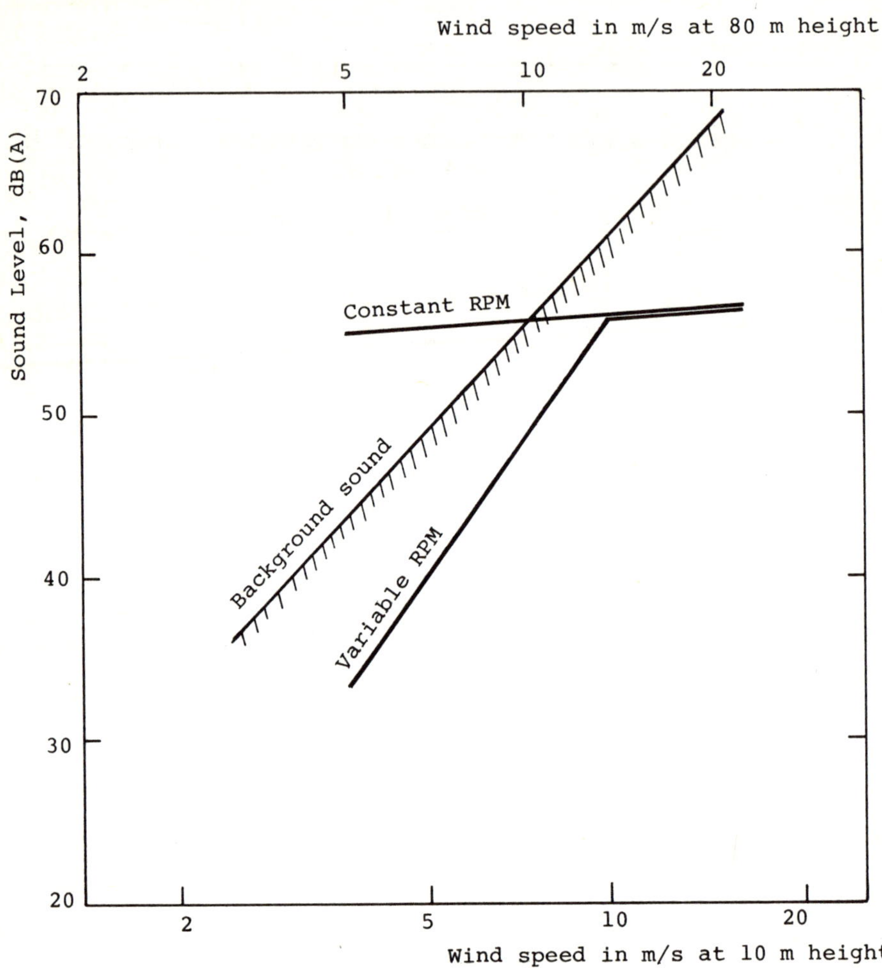

Figure 1. Calculated wind turbine noise (equivalent sound level) at a distance of 300 m from a unit of the Näsudden type. "Constant RPM" corresponds to a blade tip speed of 100 m/s. The background sound level is valid for a place surrounded by high trees.

2. SPECTRUM SHAPE

The spectrum shape of the broadband, aerodynamic noise from a wind turbine is very similar to that of the natural wind sound, which is due to the interaction of the wind with trees, shrubs, buildings etc, see Figure 2. This is hardly surprising since the noise is generated in similar ways.

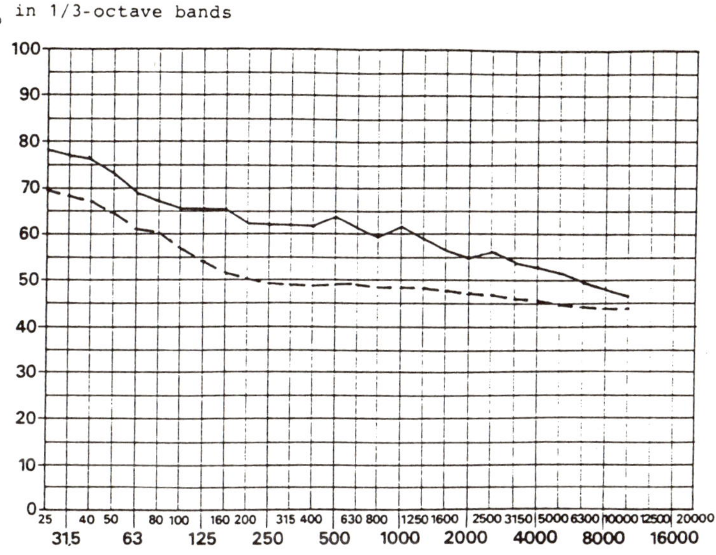

Figure 2. ------ Turbine noise, - - - - background sound. Results from Maglarp at a point 60 m downstream of the turbine.

3. NOISE LEVEL AS A FUNCTION OF ROTOR SPEED

It can be expected from a theoretical point of view that the dependence of the noise level on the rotor RPM should have the form 50*log(RPM). In order to check this relationship, measurements were carried out at a point 120 m downwind the Maglarp unit. The meas-ured values fall close to the expected line as can be seen in Fig-ure 3. However, it should be noted that the results were obtained under fairly special conditions where the generator was not con nected to the grid. For the sake of comparison, the noise level obtained during normal operating conditions and approximately the same wind speed (11-13 m/s at 10 m height) is also presented in the Figure.

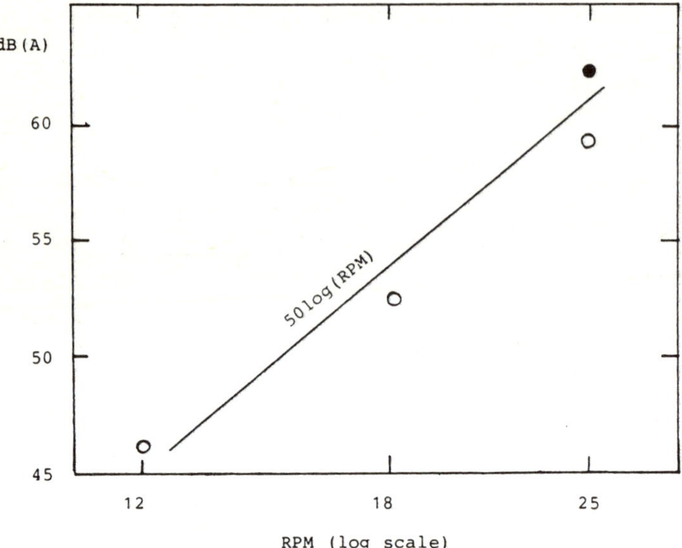

Figure 3. Measured levels at Maglarp, 114 m downstream of the turbine. ● Generator connected to the grid, o generator not connected.

4. AMPLITUDE MODULATION

A periodic amplitude modulation of the rotor noise can clearly be heard close to a wind turbine. This modulation is thought to be an important factor for the discrimination of turbine noise in the presence of natural wind sound. Results obtained at the Näsudden unit are presented in Figures 4-5. These Figures show the spectrum of the A-weighted signal (time constant: "fast"). One can see that the periodic component is small in both cases but more significant close to the source than farther away (1.0 dB at 114 m compared with 0.5 dB at 456 m). Also, an analysis of spectra obtained from different time records shows that the non-periodic modulation is more important at longer distances (reference [1]).

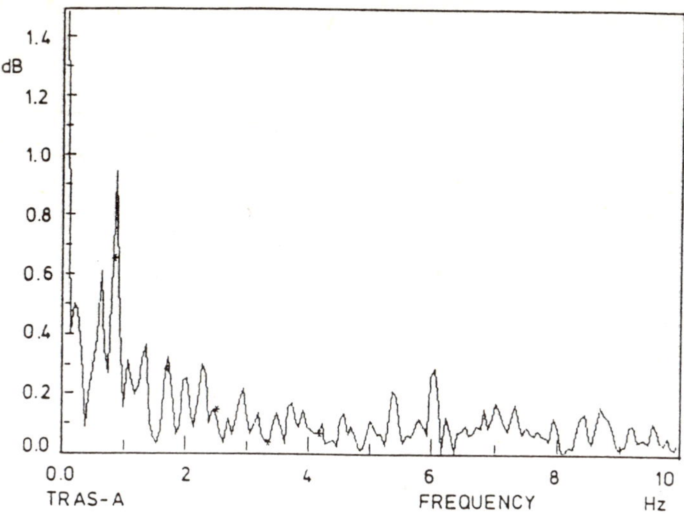

Figure 4. Modulation spectrum of the aerodynamic noise in dB re L_{Aq}, obtained at a point 114 m downstream of the Näsudden turbine. Wind speed 7 m/s at 10 m height.

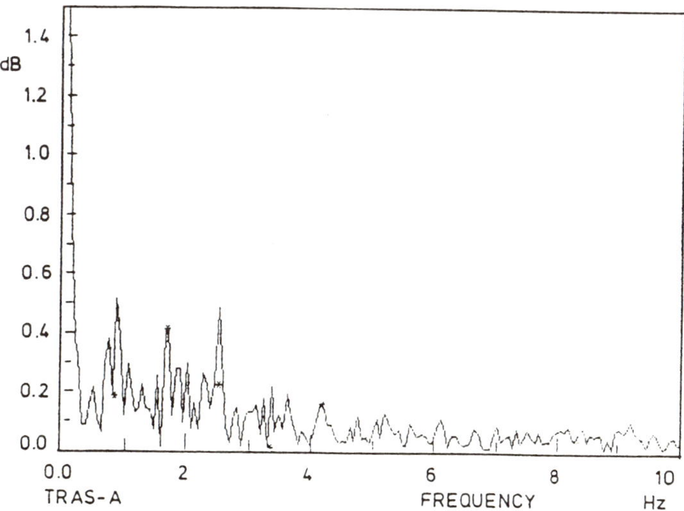

Figure 5. The same as Figure 4 but at a distance 456 m downstream of the turbine.

5. STUDIES OF BACKGROUND SOUND

The gaps in the time history of the A-weighted sound pressure level have been studied for different sites. The results are presented as a relative duration, $\Sigma T_p/T_o$, for a number of fixed levels L_{an}. T_o is the total analysing time (606 s) and ΣT_p is the total time where the level is below the chosen level L_{an}, provided the duration of each gap T_p is longer than a specified time interval. The time intervals used here are 1, 2 and 5 seconds.

The idea behind this type of analysis can be explained in the following way. Suppose that the level of the turbine noise is approximately the same as that of the background sound, and that masking in the steady-state occurs when the two levels are equal. If the level of the turbine noise, L_T, is constant in time and the level of the background sound, L_B, is fluctuating slowly, the turbine noise will be recognized during the time intervals when $L_T - L_B > 0$. On the other hand, it is believed that the turbine noise will not be recognized as such if the periods when $LT - L_B > 0$ are very short. Little is known about the relationship between the "noise-to-sound" ratio and the time needed for recognition for the small noise-to-sound ratios of interest here. Time gaps in the region of 1 to 5 seconds have been chosen as a first step.

For a measurement point near the Näsudden unit the following durations were obtained.

Duration	L_{Aq} - 0.5	L_{Aq} - 1.0	L_{Aq} - 1.5	dB(A)
› 1.0 s	13.7%	0.3%	‹0.1%	
› 2.0 s	0.12%	‹0.01%	‹0.01%	
› 5.0 s	0.08%	‹0.01%	‹0.01%	

The result shows that the time gaps cannot be considered as long and deep in the present aspect. However, the result obtained from a similar measurement at a place far away from the coast is quite different (ref [1]),

Duration	L_{Aq} - 0.5	L_{Aq} - 1.0	L_{Aq} - 1.5	L_{Aq} - 2.0	dB(A)
› 1.0 s	34%	22%	9%	2%	
› 2.0 s	33%	20%	9%	2%	
› 5.0 s	29%	18%	8%	2%	

The long-time fluctuations are obviously much more important in the latter case. The differences show that the usual cumulative distribution curves should be used with caution in this context.

6. REFERENCE

[1] S Ljunggren: Analysis of the Environmental Noise Situation around a Large Swedish Prototype Wind Turbine. The Aeronautical Research Institute of Sweden, FFA, TN 1988-16.

Chapter 8

Noise emission from a 25 metre vertical axis wind turbine—its measurement and assessment

B. L. Elliot

SYNOPSIS Detailed noise-emission measurements have been made on the prototype R McAlpine's Vertical Axis Wind Turbine (VAWT) at Carmarthern Bay under a variety of wind and operational conditions. This paper describes the methods of measurement and the analysis required to derive the basic data for siting criteria purposes and to permit comparisons with other machines.

1. INTRODUCTION

The noise emission that characterises the prototype R McAlpine's Vertical Axis Wind Turbine (VAWT) at Carmarthern Bay CEGB site has been monitored under a variety of wind and operational conditions. Methods of measurement and analysis were devised to provide basic data for siting criteria to be obtained for this machine and to permit comparisons to be made with different wind turbine generator types.

The International Energy Agency recommendations (1st Edition) on measuring wind turbine noise have been followed; these include overall noise levels, spectral characteristics, directivity and propagation.

The method of measurement using an in-house designed automatic noise sampling system (Pandora) which digitises the sampled noise and wind data and records it on tape for subsequent analysis is described. Analogue noise data is also recorded with the system for spectral analysis.

The analysis results are stated in statistical terms due to the irregular nature of the emitted noise and background wind noise. The perceived and subjective effect of the characteristic sound of the wind turbine is considered.

2. MEASUREMENT METHODOLOGY

The measuring system is based on a microprocessor controlled sampling device that accepts up to 8 inputs of noise or other related parameters (known as PANDORA, and developed by the CEGB (1)). The rates of sampling and sampling duration can be programmed or be initiated by one or more of the input signal levels (for example by wind speed or wind turbine load).

In the configuration at Carmarthen Bay, windspeed and wind-direction sensors and two Bruel and Kjaer outdoor microphones, are permanantly installed at a reference position of 50 metres from the VAWT and connected to the sampling device. Data is collected when the machine is operating and recorded on a 2-channel tape recorder for subsequent laboratory analysis. The basic set-up is shown in Figure 1. The two microphones are located at N and S compass points at

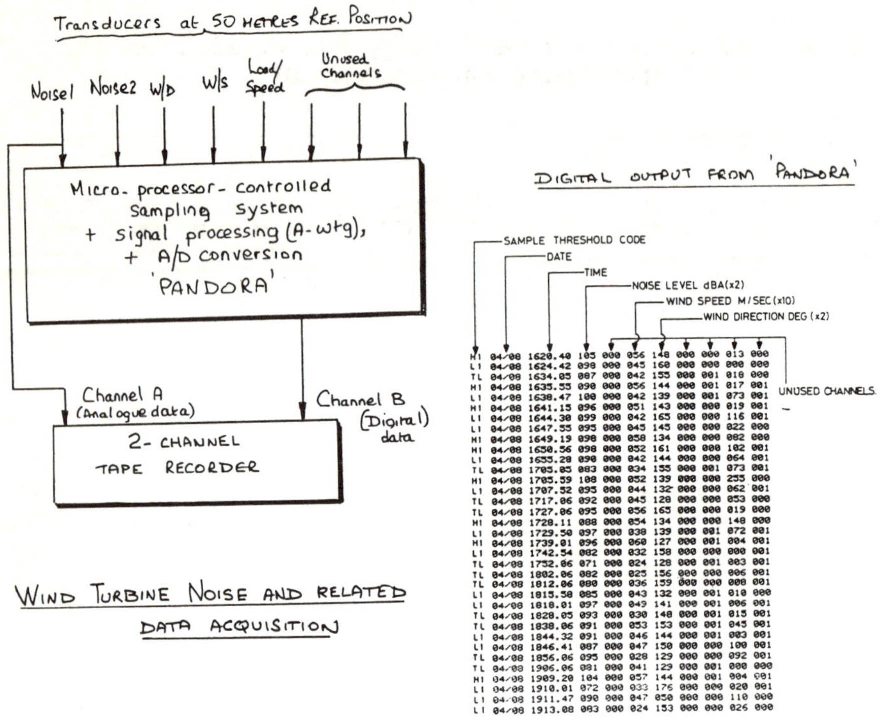

FIGURE 1. DATA ACQUISITION OF WIND TURBINE NOISE

a height of 2 metres above ground. The wind sensors are adjacent to and at the same height as one of the microphones. By monitoring the noise-emission in this way, the one permanent, unattended system (other than for the changing of tapes) enables much of the information specified in the IEA draft recommendations for wind turbine noise measurement and the proposed EEC document for same, to be obtained. The sampling rate was one every 10 seconds when the VAWT was on load and hence a considerable amount of sampled data was rapidly acquired. Each sample contained time, windspeed and wind direction and two channels of noise information. Other channels were used as required, for example, VAWT load or hub-height wind speed.

Statistical analysis of this data was performed using in-house designed software running on a PDP 11/84 computer (2). Spectral analysis was carried out on a Bruel and Kjaer FFT analyser.

3. VAWT NOISE LEVELS AT REFERENCE POSITION

Figure 2 shows the relation between VAWT noise and windspeed. The Noise parameters plotted are L_{95} dB(A) for the two rotational speeds but L_N dB for any 1/3 octave band could have been depicted with the value of N as selected. The background level/windspeed relationship is included in this figure; clearly the biggest increase in the site noise levels occurs when the VAWT is operating at low windspeeds.

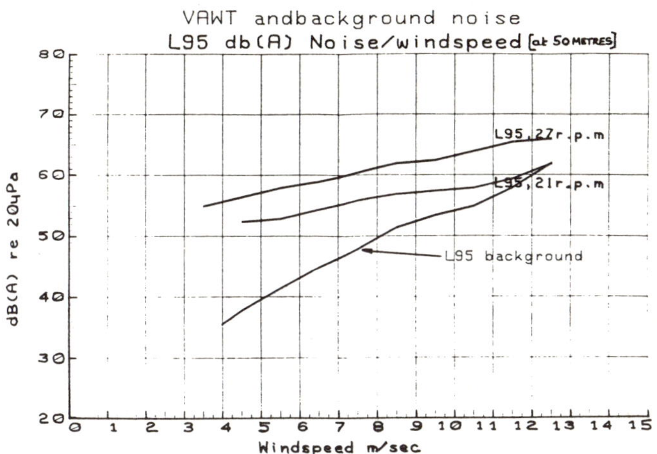

FIGURE 2. WIND TURBINE NOISE AND BACKGROUND NOISE WITH WINDSPEED

Figure 3 shows a typical time history of the variation in noise level when operating and show the modulating effect of blade noise as the upwind generating blade crosses the wind. The modulation occurs 2/rotation, in the case shown for 21 rpm operation every 1.4 seconds. Levels at a shut down of the machine are shown in the figure.

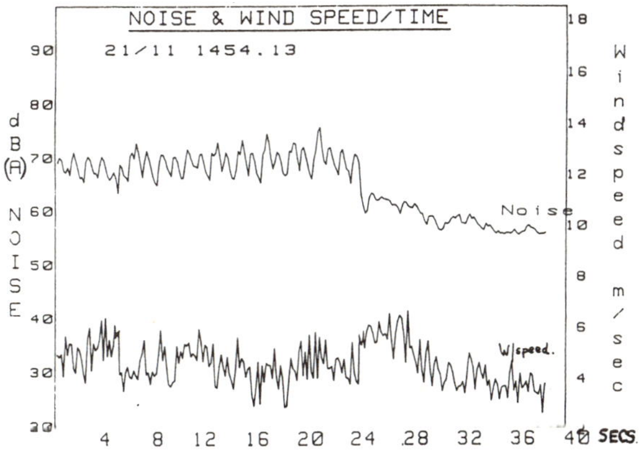

FIGURE 3. VAWT NOISE VARIATION DURING OPERATION

The directivity, if any, of the machine at 50 metres can be obtained from the same PANDORA measurements by processing the 2 microphone noise level channels with wind direction data . No directivity was established for the VAWT at this distance. Other statistical parameters easily derived from the data analysis are Leq(A) and sound power level.

4. FREQUENCY CHARACTERISATION

1/3 octave band spectra, both A- weighted and unweighted are shown in Figure 4.

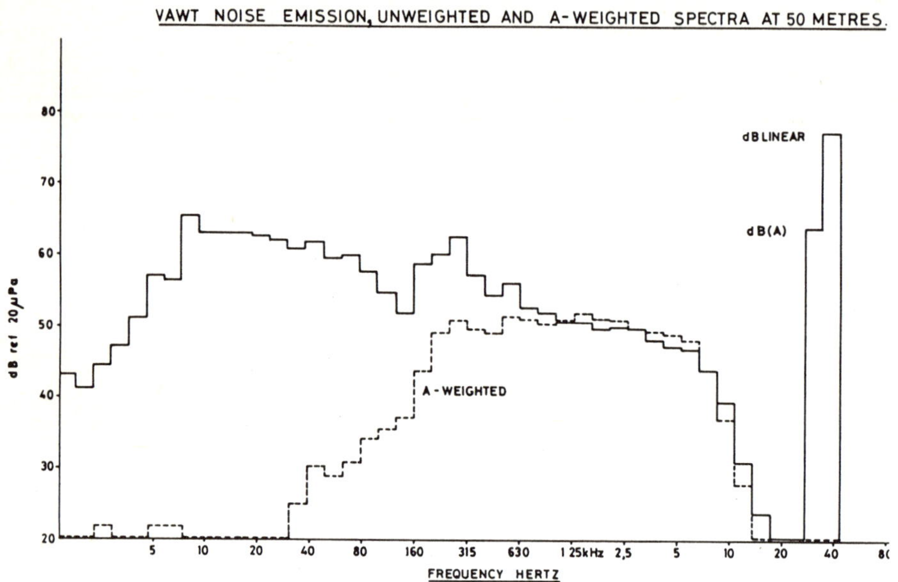

FIGURE 4. A-WEIGHTED AND LINEAR 1/3 OCTAVE SPECTRA OF VAWT NOISE EMISSION

The unweighted plot depicts more clearly the general noise emission characteristics of the VAWT namely:

(a) the broad band nature of the aerodynamic noise

(b) the increase in levels in the 160-200 Hz band of frequencies. It does not however give an indication of changing levels as the blades rotate. This is shown in Figure 5 where two 'snapshots' of frequency spectra were frozen, one subjectively assessed as corresponding to minimum noise emission and the other at maximum noise emission. A variation in level at frequencies around 170 Hz of 12dB is shown, and this characteristic feature is considered to be due to blade boundary layer/trailing noise.

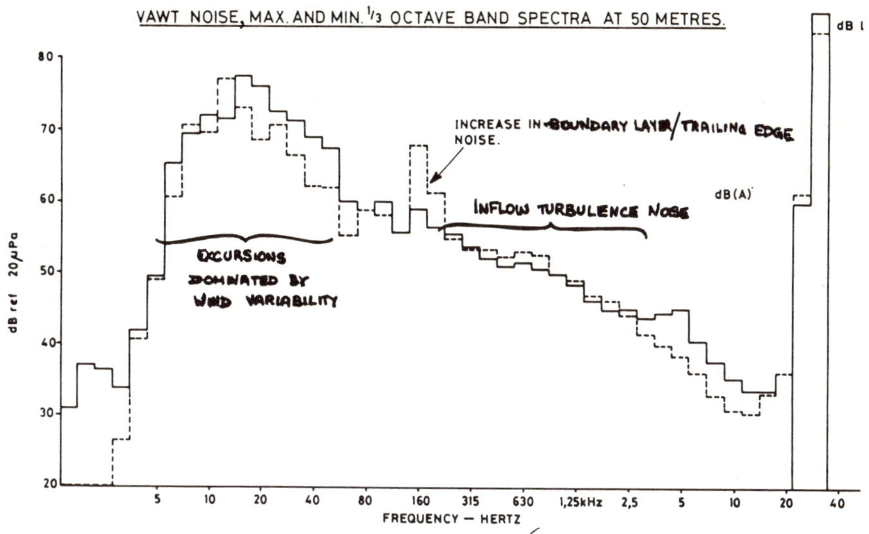

FIGURE 5. 1/3 OCTAVE SPECTRA OF VAWT MAXIMUM AND MINIMUM NOISE EMISSION

The measurements made at the reference distance were able to show noise emission other than of aerodynamic origin; Figure 6 is a narrow band analysis of that part of the frequency spectrum below 400 Hz. The aerodynamic noise centred around 170 Hz is evident, but in addition there is a tonal component at 238 Hz which has been identified as being gearbox noise.

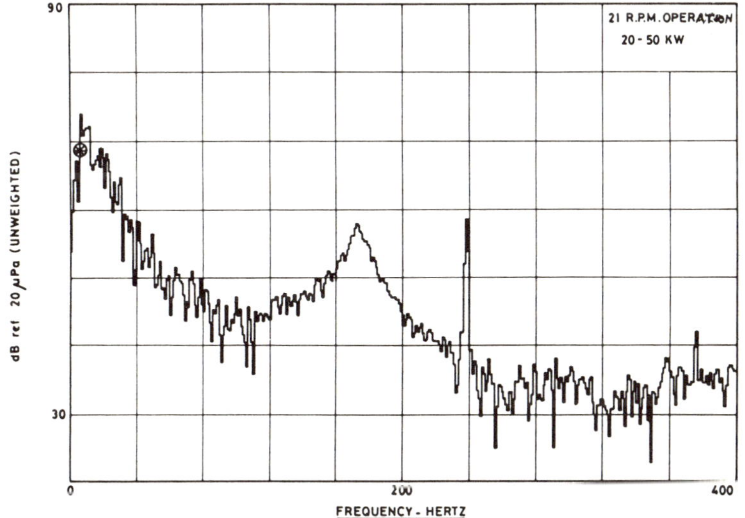

FIGURE 6. NARROW BAND SPECTRAL ANALYSIS OF VAWT NOISE

5. MEASUREMENTS BEYOND 50 METRE REFERENCE DISTANCE

To fully characterise the VAWT noise emission - for example to establish propagation - measurements were made at distances up to 600 metres from it. Figure 7 shows measurements obtained for the two operating speeds in particular wind conditions.

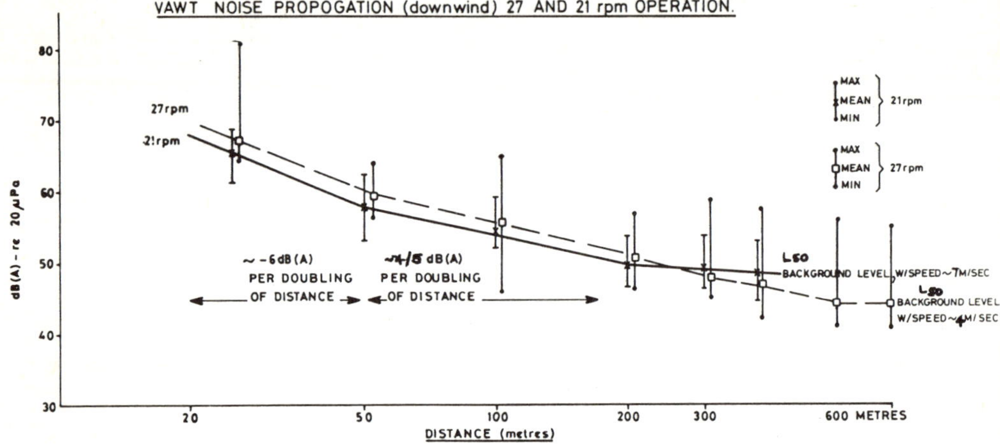

FIGURE 7. DOWNWIND PROPAGATION OF VAWT NOISE

With the propagation established and using the VAWT noise and background noise with windspeed data (eg. as in Figure 3) the audibility of the machine above the background can be assessed under different wind conditions and appropriate noise acceptance criteria be applied to the results.

6. EFFECT OF WIND ON MEASUREMENT

Finally, another factor in VAWT noise measurement that has been investigated is the effect of wind on the measurements. Figure 8 shows a dB(A) plot of noise and windspeed local to the microphone at the reference position. The short-term irregular variations in windspeed are shown, the windspeed sensor having no electrical damping. Clearly damping is necessary when relating VAWT noise to windspeed and it is decided to use a anemometer time constant of 10 seconds and with the anemometer at 2 metres above ground level to lessen the effects of windspeed fluctuations, such fluctuations are considerably smoothed out.

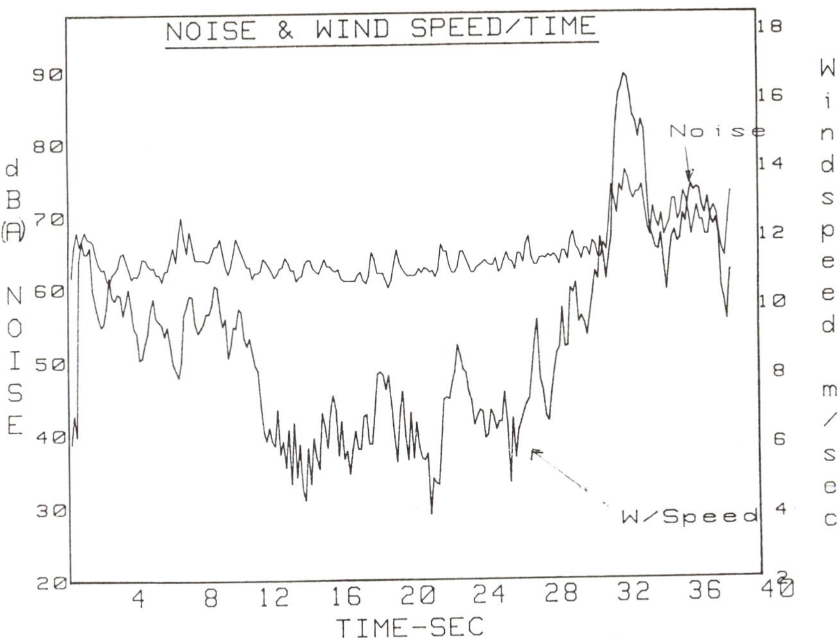

FIGURE 8. SHORT TERM FLUCTUATIONS OF WINDSPEED AND VAWT NOISE

Figure 9 shows individual measurements of noise levels obtained with the automatic sampling system, set for continuous operation with successive 10 second duration samples. The scatter indicates why a statistical approach to quantifying the measured data is desirable.

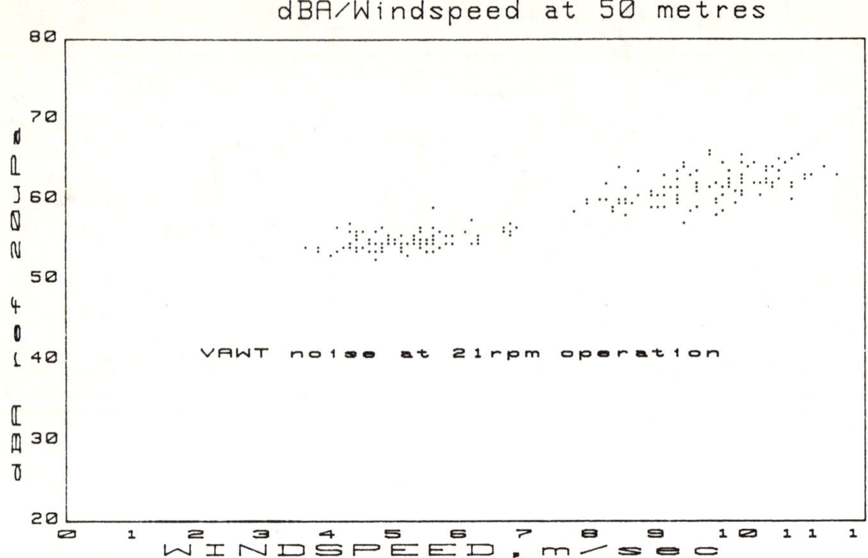

FIGURE 9. INDIVIDUAL MEASUREMENTS OF VAWT NOISE TAKEN BY AUTOMATIC SAMPLING SYSTEM

7. REFERENCES

(1) P J Nairne and M R Burt, A Programmable Automatic Noise Data and Related Parameter Acquisition System.
CEGB Report SWR/SSD/0381/N/84.

(2) B L Elliot, The Analysis of PANDORA Data.
CEGB Report SWR/SSD/0493/N/85.

DYNBE0101098/04D

Chapter 9

Environmental acceptability of the advanced wind energy converter

M. S. Vasudevan, A. Mewburn-Crook, P. R. Bullen and V. Kuhn

SYNOPSIS This paper presents results of measurements made in the vicinity of the advanced wind energy converter with the purpose of quantifying its contribution to the noise environment. Data is presented to establish in detail the spectral content of the acoustic signature of the machine. Additionally, quantification of the ambient noise environment is also undertaken, thus enabling an assessment of the environmental acceptability of the machine to be made from a noise point of view.

1. INTRODUCTION

A novel wind energy converter has been designed and developed at Kingston Polytechnic over the last three years (1). It is a vertical axis machine of high torque and low tip speed ratio with both horizontal and vertical blades rotating in an omnidirectional shroud comprising stators and a dome.

A prototype machine based on the initial development work has been built by Balfour Beatty Power Construction Ltd and Brush Electrical Machines Ltd and has been operational at Carmarthen Bay over the last two years (2). It has functioned according to expectations in respect of performance, reliability and safety.

This paper presents - for the very first time - results of measurements made in the vicinity of the prototype machine with the purpose of evaluating its acoustical impact in the light of the ambient noise environment.

2. THE ADVANCED WIND ENERGY CONVERTER (AWEC)

The original design concepts were based on operating characteristics of low tip speed ratio in order to minimise centrifugally induced stresses in the rotor and to obviate the need for active speed limiting control systems (1). This gives a machine of inherently high reliability. The criterion of insensitivity to power losses due to blade imperfections led to the design of a high solidity rotor and omnidirectional shroud, which augments the power output from the rotor irrespective of wind direction. The turbine is self starting with a torque sufficiently large to start under load.

The moment of inertia of the rotor dampens out speed fluctuations caused by unsteady wind conditions.

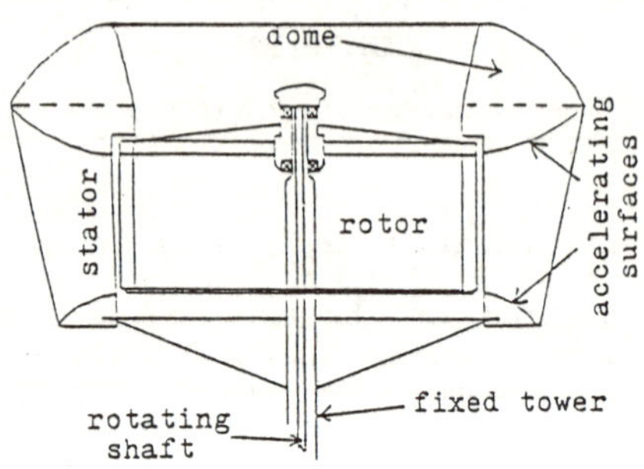

Figure 1

Figure 1 above shows the component parts of the novel augmented vertical axis wind turbine. Wind tunnel tests at Kingston Polytechnic have shown that

i) Each component part makes a significant contribution to the performance and that the augmentor improves the peak power coefficient by a factor of 2.5.

ii) The power coefficient is high at a tip speed ratio of 1.3 and a run away tip speed ratio of just under 2.0.

iii) The overall dimensions of the wind turbine, including the augmentor, are appreciably smaller than equivalent existing machines - thus optimising its cost effectiveness.

3. INSTRUMENTATION AND DATA ANALYSIS

Noise measurements were made by employing a standard combination of Bruel & Kjaer condensor microphone (type 4145) and Nagra IVSJ tape recorder. In all cases analogue signals were recorded on the tape recorder for detailed spectral analysis on a Rion SA-73 dual channel sound and vibration analyser. A schematic of the instrumentation chain is shown in the figure below. The system was calibrated before and after use with a B & K calibrator (type 4230).

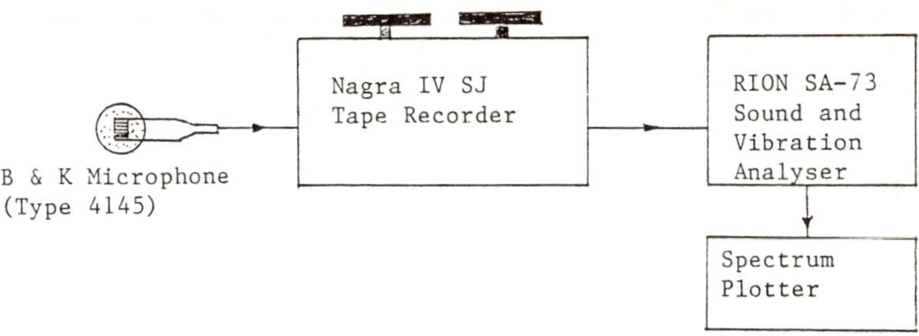

Figure 2 Schematic of chain for Instrumentation and Data Analysis

Throughout care has been taken to check the electronic noise floor of the system and to ensure the measurement of real acoustical levels well above the electronic noise floor.

In order to understand the spectral characteristics of the acoustic signature of the prototype advanced wind energy converter as well as that of the ambient noise environment, it is necessary to carry out "high resolution" spectral analysis of the recorded signals. Such an analysis helps to distinguish wide band noise sources, if any, from narrow band sources. Analysing on a constant band width basis rather than the more usual constant percentage band width has therefore been preferred.

However, 1/3 octave spectra were sometimes obtained before deciding on the frequency range for fine resolution narrow and analysis. Parameters for spectral analysis were chosen such that an accuracy for spectral levels of ± 1dB was achieved with a confidence level of 96 - 98%.

4. FIELD MEASUREMENTS

The major concern of this work is with the acoustic signature of the Balfour Beatty machine. Measurements in its vicinity were complicated, however, by the operation of two other wind energy machines operating in close proximity (less than ½Km) and outside our control. It is essential to note here that the McAlpine machine was functioning during the entire duration of these measurements. The horizontal axis Howden machine, however, was operating intermittently. Clarification in this respect is given at the appropriate point in the discussion and in the presentation of results for noise spectra.

Measurements were made at several positions at a radial distance of approximately 45m from the Balfour Beatty machine (positions I to IV). Position I was directly downwind in a northwesterly direction and the spectrum obtained is shown in fig 4.

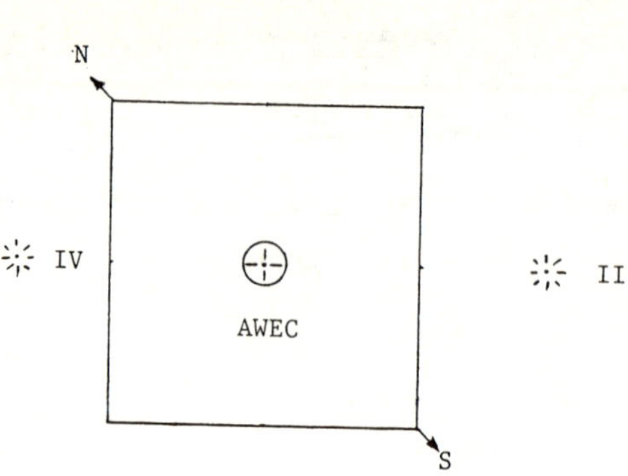

Figure 3 Measurment Locations Around The AWEC

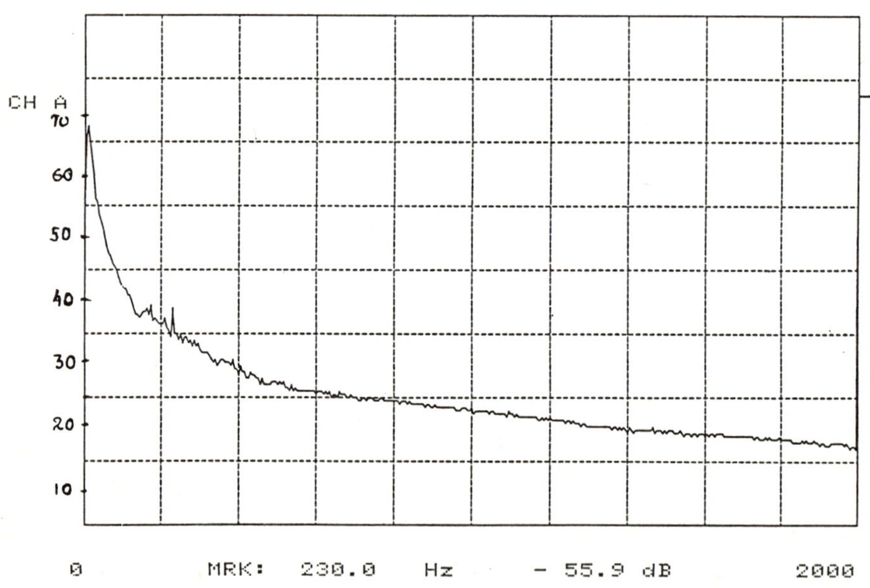

Figure 4

The spectrum is generally broad band in nature with some energy around 200Hz (<40dB). A pure tone component at 230Hz is also identifiable.

Figure 5 below shows the spectrum obtained at position III, in an upwind direction. It is broadly of unchanged character but is now generally 10 - 15dB lower than the downwind spectrum. Positions II and IV either side of the wind direction yielded spectra of identical shape and differences in spectral levels were insignificant (figures 6 and 7).

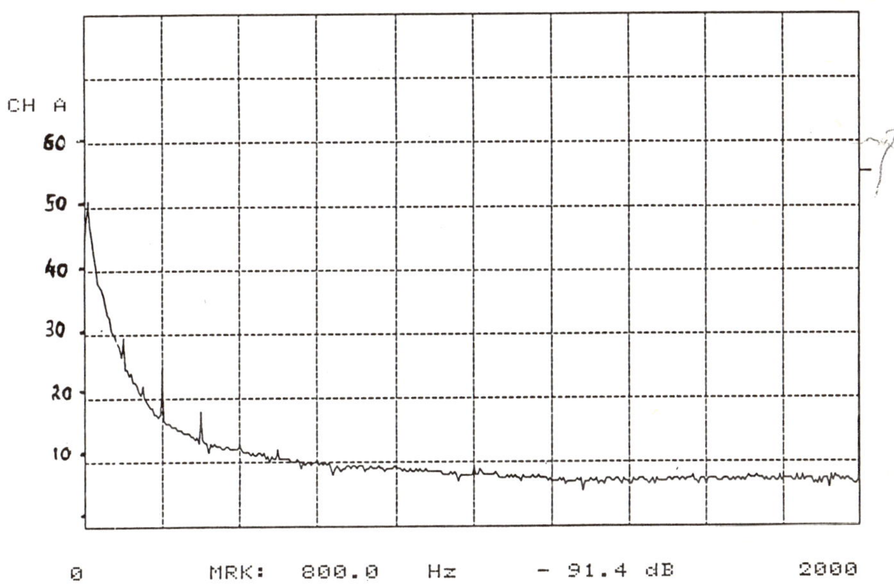

Figure 5 Spectrum for Position III, upwind

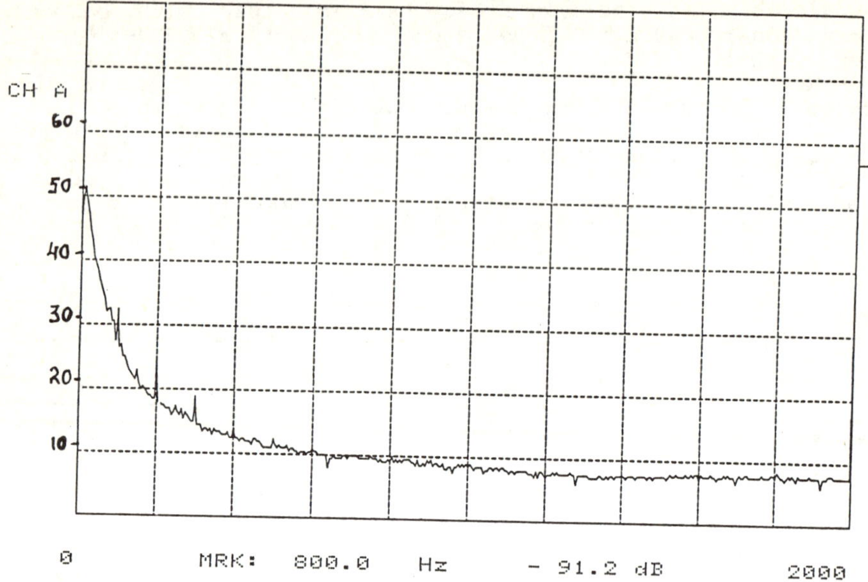

Figure 6 Spectrum obtained for position II

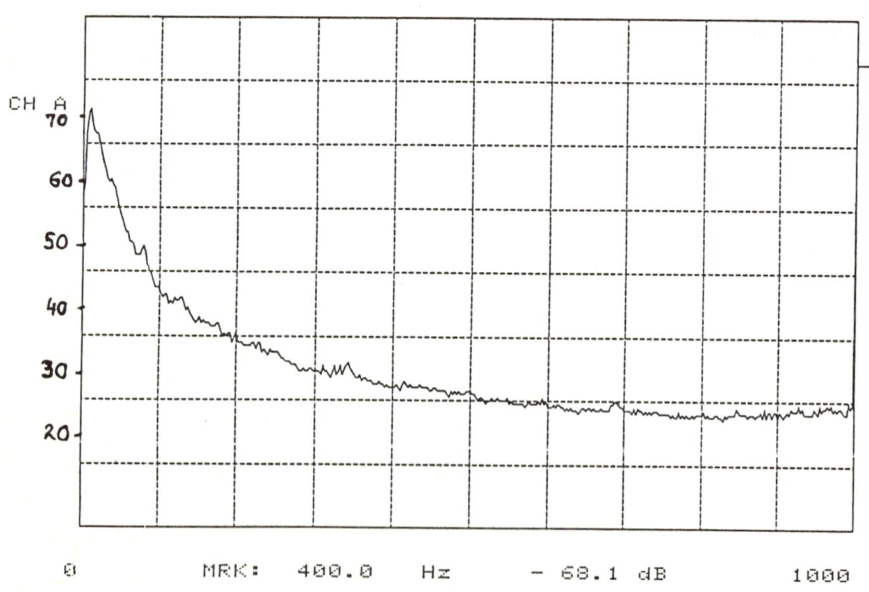

Figure 7 Spectrum obtained for position IV

During the above measurements the meteorological conditions were monitored continuously (temperature, wind speed and direction) and showed no major change. Position IV was a location where the Balfour Beatty machine was clearly audible, but spectral analysis with improved resolution (figure 7) failed to show any discrete frequency components. Energy around 80Hz is attributable from experience to distant traffic rumble.

As position I gave the highest levels, it is of interest to compare that spectrum with that of the ambient background, but with the Balfour Beatty machine stopped (figures 8 and 9).

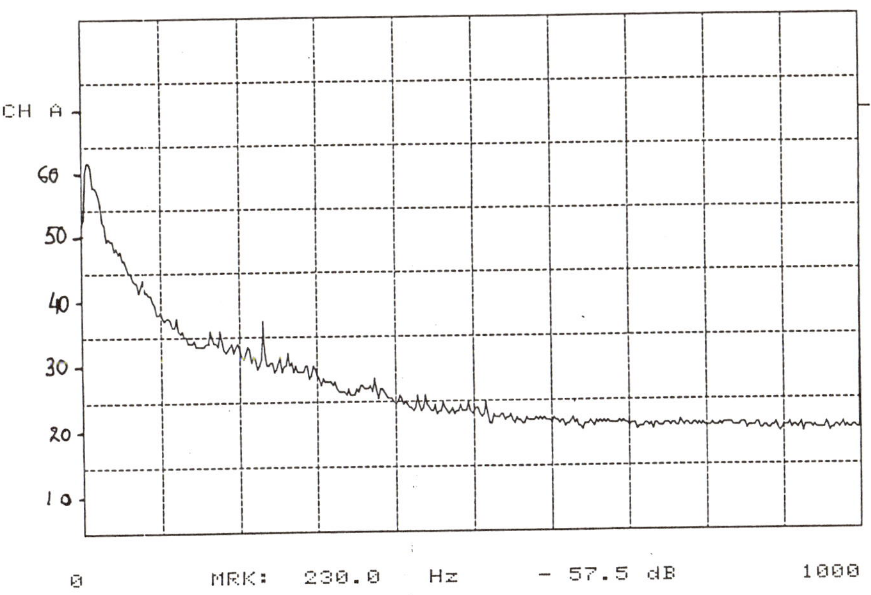

Figure 8 Spectrum obtained with AWEC on

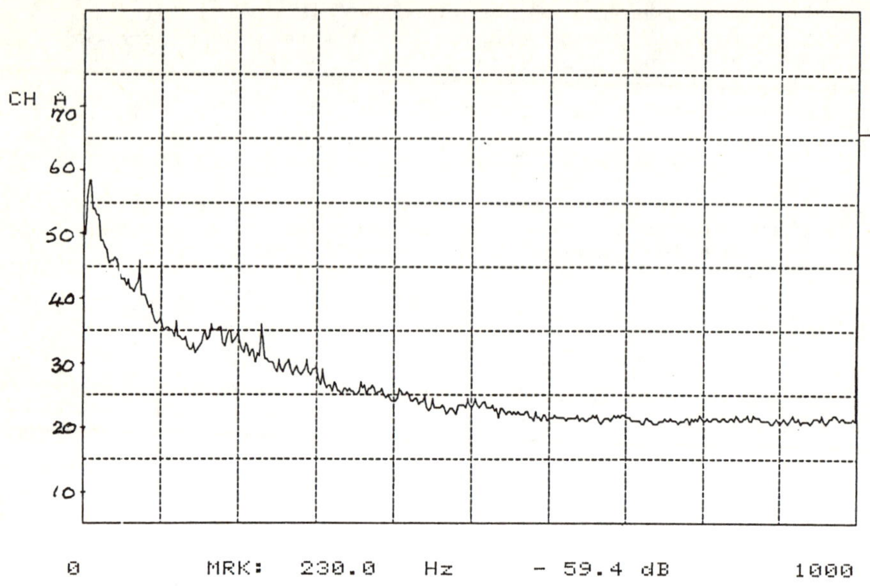

Figure 9 Spectrum obtained with AWEC off

It is interesting to see that figure 9 continues to show narrow band energy around 230Hz in addition to broad band energy around 200Hz. Clearly, therefore, the AWEC is not the cause for the above features of the spectra.

Further more, differences in level of no more than 1dB or so between figures 8 and 9 [figure 8 levels marginally higher throughout] emphasises the remarkably insignificant contribution of the AWEC to the total noise spectrum.

All the results presented thus far were obtained for measurements at various positions around the AWEC, with the McAlpine vertical axis machine contributing to the background all the time. It is interesting to consider the general noise spectrum when all the three machines are functioning (figure 10).

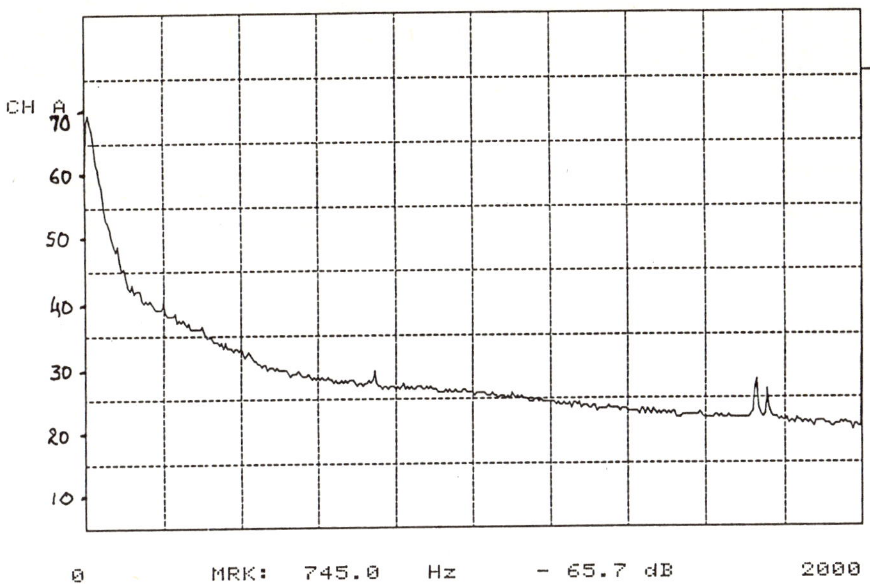

Figure 10 Spectrum obtained with all three machines operating

On comparison with figure 4, it is seen that the spectral shape and levels up to 2KHz remain largely the same except for the appearance of twin peak components at 1730Hz and 1760Hz (below 30dB). These components may well be due to the Howden horizontal axis machine.

CONCLUSION

Based on preliminary measurements, the Advanced Wind Energy Converter makes a remarkably low contribution to the noise environment.

ACKNOWLEDGEMENTS

The authors gratefully acknowledge the help rendered by Dr R N Vasudevan during the measurements on site and in analysing the data subsequently. The support of SERC, Balfour Beatty Power Construction Ltd and the DTI is also appreciated and acknowledged.

REFERENCES

BULLEN, PR; MEWBURN-CROOK, A; and READ, S. 'The engineering and economics of an augmentor for a vertical axis wind turbine' BWEA Annual Conference, London, March 1988.

BULLEN, PR; MEWBURN-CROOK, A; and READ, S. 'The performance of a novel shrouded vertical axis wind turbine' BWEA Annual Conference, Edinburgh, April 1987.

Chapter 10

Electromagnetic interference—the requirements of a method of machine testing and approval

R. J. Chignell

SYNOPSIS The rotation of wind turbines disturbes their local radio environment and produces interference. All types of radio systems are potentially at risk but the consequences range from being life threatening, through major losses of revenue, to the disturbance of entertainment by TV broadcasting.

If wind energy is to realise its full potential then future installations will increasingly be near centres of population and more radio services will be affected. The objective of this paper is to propose a means of testing wind turbines that will enable more effective guidelines aimed at limiting interference to be established. It is proposed that if developed these could be applied to each type of machine rather than each site.

1. INTRODUCTION

The objective of this paper is to indicate how progess may be made towards establishing test criterion for the acceptance of wind machines adjacent to existing radio installations.

The IEA Expert Group(1) concerned with Electromagnetic Interference from wind turbine generators recognised that although disturbance of TV broadcasting was the most widely reported problem it was neither life threatening nor incurred the heavy financial penalty associated with interference to other radio services. For example, major concern was expressed about the aeronautical radio navigational services at VHF where disturbance could result in a passenger aircraft being mis-routed. Existing regulations prevent the erection of fixed structures with moderate dimensions within 1 Km of a VOR station (VHF Omnidirectional and Ranging) and longer distances apply for larger structures. The rotating structure of a wind turbine could generate other problems.

Concern about interference to microwave links has been expressed by two European countries, Denmark and Netherlands. The value of daily traffic carried by some microwave links is significant in terms of the capital costs of the wind installations concerned. These are two examples of radar services where serious disruption can result from electromagnetic interference arising from wind generators. The IEA Expert Group specifically considered nine types of radio services, but there are many other services that should be included.

The radio community has a long history of international self-regulation to prevent interference problems. CISPR (International Special Committee on Radio Interference) was established in 1933. Every new radio service has been the subject of extensive scrutiny and planning to prevent interference to established services and optimise the use of the radio spectrum, possibly nature's most scarce resource. There is also a great weight of established regulations and requirements, backed by law, which can be directed at any perceived problem.

It is into this background that new wind machines must be introduced, and if wind energy is to achieve its potential then there will be occasions when turbines will be competing for prime hill top sites with established radio systems. So far many of these problems have been avoided because large wind machines have been installed in remote locations, for example, in the UK, Orkney and Carmarthen Bay, are not close to large centres of population. Proper exploitation of the wind resources cannot be accomplished without procedures that allow turbines to be introduced in a harmonious manner reasonably close to established radio systems. Minor changes to existing radio requirements could for example prevent the erection of a large wind turbine within 5 Km of a VHF aeronautical installation. There are over 500 such installations in the UK and such a condition would sterilise approximately 40,000 sq Km for wind generation.

The objective of this paper is to indicate some of the avenues that should be pursued in order to treat the problem of electromagnetic interference in a manner that will both obtain acceptance in the radio community and allow realistic wind turbine installation procedures to be adopted.

2. THE ORIGIN OF INTERFERENCE

Wind turbines create electromagnetic interference to radio services by generating time varying multipath effects. The presence of a wind turbine introduces a second transmission path linking the radio receiver with the transmitter. Multipath is a common feature of radio propagation. Large buildings, trees and other structures generate multipath but it is the time variation of the multipath effects introduced by wind turbines that is so destructive to radio systems. The time constants of the rotating wind turbine blades are comparable to those of the radios' automatic control circuits and radios are frequently vulnerable to such effects.

There are many different electromagnetic scattering mechanisms present on a wind turbine. The blades of the machine are the rotating components and will play the major role in generating electromagnetic interference, but they may interact with the support tower either by chopping the energy scattered from it or by supporting a combined resonant current distribution. Given the frequency range of radio signals to be considered their variety of modulation schemes and the wide variation in the parameters of wind turbines many factors are important in some circumstances. The parameters of the wind turbine that need to be considered have been discussed elsewhere(2) but include:

- (i) the type of machine
- (ii) the machine dimensions
- (iii) the rotation speed
- (iv) the blade construction
- (v) the blade angle and geometry
- (vi) the support tower

3. MACHINE MANUFACTURERS REQUIREMENTS

From the discussion presented so far the potential complexity of the electromagnetic interference problem can be seen. There are many radio systems potentially at risk, there are many parameters of the wind turbine that in particular circumstances could become critical and given the range of possibilities there is little experience of practical problems. In this situation the IEA Expert Group produced Preparatory Information to attempt to identify situations where a problem may exist. On a site by site basis this essentially recommends a survey of the transmissions actually present, together with an investigation of the potential vulnerable receivers within the interference zones.

In remote areas any site work is expensive although the number of transmissions that have to be investigated may be few. In more urban areas the number of radio services involved will be greater and the number of potentially vulnerable receivers will increase greatly. For example, the planning assumption of one television per household would not be unreasonable and the expense of providing alternative cable TV services could again be considerable.

The present preliminary information may prevent sites with the worst electromagnetic interference environments from being chosen, but they neither aid a manufacturer with the next installation, nor provide information on the key to the whole problem, of how to design a machine that minimises the electromagnetic interference. A machine that was invisible to radio wave would not generate electromagnetic interference, but the costs associated with some of the military "stealth" programmes indicate that this may be an unrealistic objective.

In the absence of such a solution, what manufacturers require is the ability to test their machine once and know its performance, i.e. that it can then be successfully operated with circles of known diameter enclosing the interference zone at specified frequencies and for particular types of radio systems. The remainder of this paper will be devoted to indicating the steps necessary to work towards such a goal.

4. THE WAY FORWARD

Given the wide range of possible problems that could be encountered and the lack of extensive experience, except perhaps in the area of television broadcasting, it would be straightforward to propose an extensive electromagnetic interference testing and measurement programme for all type of wind turbine. However, given the range of problems to be spanned the test programme could be long and laborious. The electromagnetic resonant effects of wind turbines vary rapidly with frequency and the modulation properties of radio transmissions in adjacent band throughout the wide electromagnetic spectrum are very different. It is evident that a limited measurement programme must be undertaken but that it should span the range of problems. It is reconciling these two different pressures that has prevented effective progress in establishing useful guidelines for machine manufacturers.

The key to unlocking this problem is in recognising that there are many essentially broadband electromagnetic scattering mechanisms present at a wind turbine and it is only when they are combined that narrowband effects occur. Thus the way forward is to undertake measurements in a way that enables the different contributions to be separated, individually characterised and their methods of combining understood. In this way limited sets of measurements on a machine would enable much more information to be computer generated, on its properties over a more extensive bandwidth and range of condition.

Worst case conditions could then be unambiguously identified in terms of a particular type of machine interfering with specified types of radio services. More importantly the most significant scattering mechanism for a particular machine could be identified, allowing remedial action to be taken. The scope for design changes will depend upon which mechanism is dominant, it may be scattering from the leading edge of a blade, modulation of scattering by the tower or from the blade root. Some sources of interference will be easier to control than others but until the detailed mechanisms are more fully understood there is no scope for making design improvements to reduce electromagnetic interference from one generation of wind machine to the next.

5. THE REQUIREMENTS OF A MEASUREMENT SYSTEM

The characteristics of the different electromagnetic scattering mechanisms present on a wind turbine, essentially requires an unusual type of short range radar. The wind turbine must be illuminated by a suitable radio transmission and all the scattering components separated. Essentially this means they must be separated in order of their time of arrival and the times related to the instantaneous wind machine geometry.

The analysis of such data must be undertaken in conjunction with an electromagnetic scattering model of the wind turbine that enables the different measured components to be identified and combined in different ways to simulate changes in key parameters such as frequency. Such models are available in other applications of electromagnetics and similar principles could readily be applied to wind turbines.

The other key aspect of the problem is the effect the signals generated by the wind turbine have upon the potentially vulnerable radio systems. Based upon the analysis of the radar type measurements the ranges of waveforms a wind turbine can generate may be synthesized. Such waveforms can now, via digital to analogue convertors be synthesized in analogue form and be injected into a radio system to simulate the field conditions that would be observed around a wind turbine.

All the hardware described above could have been made in the past, but recently all the important sub-systems have become available as off-the-shelf commercially available items. This is an important step as with suitable software back up in principle it allows all wind operators to undertake the sophisticated measurements that are necessary to understand and reduce electromagnetic interference effects.

It is proposed that the first step in this process is to introduce such measurements at established wind turbine test sites. From that base the techniques could be more widely applied and a library of observed effects on actual machines established.

6. CONCLUSION

In this paper, a way forward to the better understanding of electromagnetic interference effects caused by wind turbines is proposed. It is based upon the use of commercial instrumentation that has only recently become available and upon the realisation that although some scattering effects appear to be narrowband, they are caused through the combination of a number of essentially wide band contributions. By characterising the individual contributions and the way they combine from a limited range of measurements, much more widely applicable conclusions may be drawn. Investigations of the individual contributions will also provide data for machine designers that will enable improvements to be made in subsequent generations of machine. This is the route to the effective reduction of electromagnetic interference from wind turbines.

7. REFERENCES

(1) Recommended Practices for Wind Turbine Testing 5 Electromagnetic Interference Preparatory Information Issue 1, February 1986, Edited by R J Chignell

(2) R J Chignell, Electromagnetic Interference from Wind Energy Conversion Systems - Preliminary Information, EWEC 86, 7 - 9 October 1986, pp 583-586.

Chapter 11

Large wind turbines; a source of interference for TV broadcast reception

P. J. van Kats and J. van Rees

SYNOPSIS

A planning-model for site allocation of large windturbines (Wind Energy Conversion Systems, WECS) with respect to their impact on radio services is proposed.
This model is based on measurements of the radar-cross-section of WECS. These measurements have been performed at several locations around the 45 m WECS near Medemblik in the Northern part of the Netherlands.
From the results, the signal-to-interference ratio (C/I); for UHF TV broadcast reception in the area around the WECS was calculated and compared to the TV picture quality.

1. INTRODUCTION

The development of wind energy as an alternative source of energy is characterized by the application of Wind Energy Conversion Systems (WECS) with an increase of installed power. The resulting increase in size of these WECS will seriously affect the hazard of interference to radio services.
Reflection as well as scattering and diffraction can affect the quality of the radio path. Besides, operational WECS will also cause a periodic disturbance due to the movement of the rotorblades.
However, to what extent these interfering signals will affect the quality of a radio service, strongly depends on the type of radio service (e.g. mobile radio, TV broadcast, line of sight links, radar).
Earlier work [1] reveals that especially TV signals are vulnerable to this type of periodic interference, as the video content of the signal is also amplitude modulated.
Especially for TV broadcast where a large area is served, the site allocation for the erection of large WECS has to include an evaluation of TV interference risks.

This study was performed by the PTT, Dr. Neher Laboratories, under contract with the Management Office for Energy Research (PEO), Netherlands

2. INTERFERENCE MECHANISM

As mentioned in the introduction, a WECS behaves as an obstacle for incident electromagnetic waves. Reflection, scattering and diffraction phenomena will spread the incident electromagnetic field over its environment; thus the WECS can be considered as a secondary source of radiation.
The general geometry is shown in figure 1.

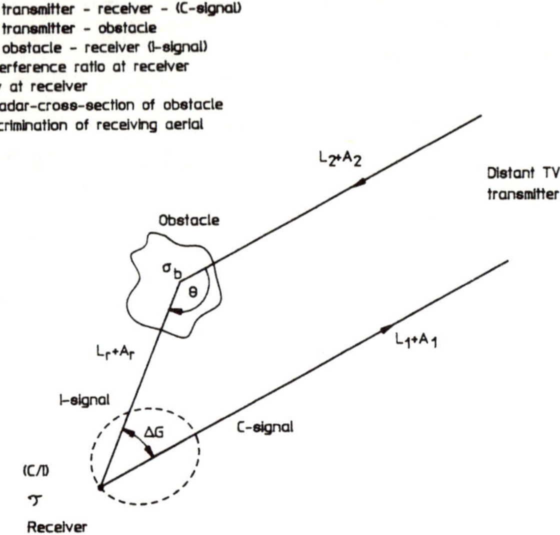

Figure 1: Geometry for interference on the radio path by an obstacle

The extent to which a WECS spreads this secondary field over its environment is largely determined by the following factors:

- size and shape of the obstacle with respect to the signal wave-length λ;
- dielectric and conductive properties of (parts of) the WECS,
- position of the blades and the structure of the WECS with respect to the polarisation of the incident wave.

As a WECS in operation is in continuous motion (especially its rotor-blades) the impact of the reflected and scattered field shows an unpredictable behaviour.
Calculation of the scattering and reflecting properties of these complex bodies in a changing geometry is complicated, thus additional measurements will improve knowledge on the behaviour of WECS on radio waves.

3. BASIC EQUATIONS FOR THE PLANNING MODEL

In RADAR technology, practical use is made of the impact of an obstacle on radio waves. From this, an obstacle is characterized by its (bistatic) radar-cross-section (r.c.s): σ_b. σ_b is the imaginary surface of an isotropic radiator from which the received power corresponds to the actual power received from the obstacle.

In general σ_b is a function of (di)electric properties and geometry of the obstacle and of the signal wavelength.

When the impact of large WECS on radio waves is translated into its bistatic r.c.s. an expression for the signal-to-interference ratio (C/I) can be found in the area around the WECS.

For any specific radio service (e.g. TV broadcast, mobile radio, microwave links) however, a different (C/I) ratio will be required for reliable operation.

Under the assumption of the geometry shown in figure 1 the (C/I) at a distance r from the windturbine can be calculated as follows:

Wanted signal (C):

$$C = P_t - (L_1 + A_1) + G_r \qquad [\text{dBW}] \qquad (3\text{-}1)$$

Unwanted signal (I):

$$I = P_t - (L_2 + A_2) + 10 \log \frac{4\pi\sigma_b}{\lambda^2} - (L_r + A_r) + G_r' \qquad [\text{dBW}] \qquad (3\text{-}2)$$

$L_1 = L_2$ when the distance of the WECS and the receiver to the TV transmitter is large.

The free space loss L_r can be calculated from:

$$L_r = 20 \log 4\pi + 20 \log r - 20 \log \lambda \qquad [\text{dB}] \qquad (3\text{-}3)$$

Introducing the antenna discrimination factor

$$\Delta G = G_r - G_r' \qquad [\text{dB}]$$

then the (C/I) becomes

$$(C/I) = 10 \log 4\pi + 20 \log r - 10 \log \sigma_b + A_2 - A_1 + A_r + \Delta G \qquad [\text{dB}] (3\text{-}4)$$

From this expression follows that (C/I) can be improved by:

* decreasing σ_b(reducing the effect of the WECS);
* increasing ΔG (better directivity of the antenna);
* increasing additional loss $A_2 + A_r$ on interference path;
* decreasing additional loss A_1

Only a variation of these factors may help to reach the required (C/I) when interference actually occurs.

Possible control by these parameters stronly depends on the location of the receiver with respect to the WECS.

In general two alternative situations can be distinguished:

a) A significant delay time τ between wanted signal (C) and unwanted signal (I). The receiver is located in an area between the WECS and the transmitter. In this area, the unwanted signal (I) is dominated by reflection and scatter from the WECS and is called the "backscatter region".
b) No delay time between the wanted signal (C) and unwanted signal (I). This phenomenon is met on places where the receiver is behind the windturbine, as seen from the transmitter. In this area the unwanted signal can only be the result of scatter or refraction at the WECS and is called the "forwardscatter region".

In equation (3-4) the distance r from the WECS to the receiver can be considered as a variable. When the factors A_1, A_2 and A_r are assumed constant, $\Delta G = 0$ and σ_b of the windturbine is known, then $r = f(C/I)$ can be calculated.

When the (C/I) is interpreted as the (C/I) required for a specific radio service, a curve with polar coordinate: r around a windturbine can be drawn, covering an area inside which the required (C/I) will not be satisfied.

With this method, criteria on the site allocation of large WECS with respect to their interference on TV reception can be produced, when:

a) the (C/I) required for TV reception is known;
b) the bistatic radar-cross-section of the windturbine is known.

4. TV PICTURE QUALITY

As previously stated, a WECS can be considered as a secondary radiating source (of the I-signal), interfering with the direct, wanted source (the C-signal e.g. TV transmitter).
TV picture quality is dominated by the ratio of (C) and (I), thus by (C/I). The extent to which a given (C/I) is harmful for TV picture quality is different for static and dynamic (e.g. rotating) obstacles.

For dynamic (moving) obstacles two different phenomena may occur according to the geometry between TV transmitter, obstacle and TV receiver:

a) in the backscatter region, due to the time delay τ, a periodically fluctuating "ghost image" is observed;
b) in the forward scatter region, no "ghost image" can be observed, but the TV picture intensity fluctuates in accordance with the periodic movement of the obstacle due to a slow attack of the Automatic Volume Control in most TV sets.

From experiments with a simulation set-up with different TV-sets and 5 different observers (all technicians) it was concluded that the following TV picture quality scaling could give a fair approach to the TV picture degradation due to dynamic interference:

Δ [dB]	C/I [dB]	CCIR qualification	to expect
< 0.1	> 39	excellent	no interference
0.2	33	good	visible interference
0.4	27	fair	disturbance
0.8	20	bad	annoying disturbance
2.5	10	extremely bad	severe disturbance

Table 1

Recordings of the signal variation in the forward scatter region (see Appendix) show quick regular fluctuations riding on the undulating amplitude variation caused by the slow rotation of the rotor blades.
In some publications [1] this signal variation Δ is expressed in terms of the modulation index m using:

$$\Delta = 20 \log (1 + m) \quad (4\text{-}1)$$

A table showing corresponding values of (C/I), Δ and m is given below:

(C/I) [dB]	signal variation Δ [dB]	modulation index m
38.7	0.1	0.01
32.7	0.2	0.02
26.5	0.4	0.05
20	0.8	0.1
9.5	2.5	0.33
0.0	6	1

Table 2

5. MEASUREMENT RESULTS OF THE (BISTATIC) RADAR CROSS SECTION

From the equations in paragraph 3 it became evident that once the (bistatic) radar-cross-section (r.c.s.) of an obstacle is known, the polar coordinate r encircling the obstacle can be calculated for a required (C/I).
It was also noted that two different regions around the windturbine can be distinguished viz:

a) backscatter region;
b) forwardscatter region.

For both regions different measurement methods were used.

The measurement procedure was performed around the 45 m WECS near Medemblik after questions concerning TV broadcast interference, in the forward- scatter region.

A TV broadcast transmitter operating at 2 UHF frequencies:

Nederland 1 picture 615.25 MHz, sound 620,75 MHz
Nederland 2 picture 663.25 MHz, sound 668,75 MHz } horizontal polarisation

is located near Wieringen about 15 km from the location of the WECS.

Measurements were performed at 8 different spots around the windturbine (see figure 2); besides continuous TV-picture monitoring was used in the complete area around the WECS.
Movement of the WECS was established by rotation of the main support structure (milling angle: 0 deg. = west, 90 deg. = north, 180 deg. = east, 270 deg. = south) and rotation of the rotor blades under motor drive. Main position of the blades was set to normal working setting (maximum observed surface).

5.1 Backscatter

Backscatter measurements were performed at position nr. 8 (figure 2) at 950 m distance from the windturbine. Measurement results show a strong dependence on the rotating position of the blades (see results in the Appendix).

The reflected power varies from "no detectable influence" to tens of dB above the noise level. From these results the r.c.s. was estimated at 24 dBm^2

5.2 Forwardscatter

The forwardscatter measurement was conducted at seven different locations (figure 2). Applying equation (3-4) and table 2, from which the relation between the measured power variation and (C/I) can be derived, the bi-static r.c.s. was calculated.
At location nr. 2 considerable interference from the windturbine was measured. The largest variation occurs when the rotor blades are in almost horizontal position (see table 3).

location	milling angle	amplitude variation	(C/I) [dB]	σ [dBm^2]
2	240	1.7	13.4	39.1
2	250	1.9	12.2	40.3
2	260	2.2	10.8	41.7
2	310	0.8	20.0	32.5

Table 3. Amplitude variation, corresponding (C/I) and calculated bistatic r.c.s. for location nr. 2 at different milling angles.

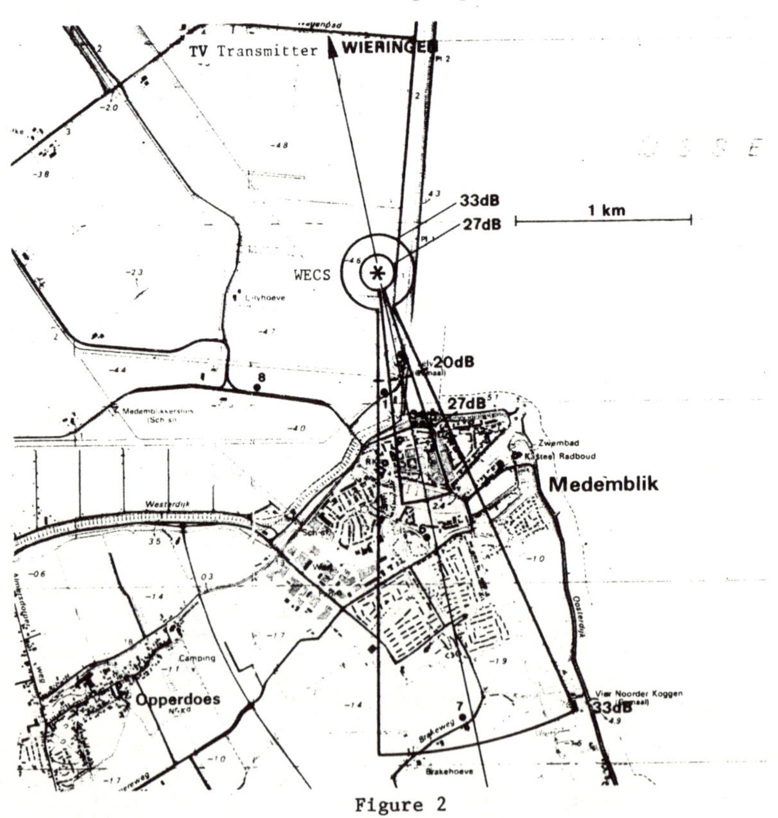

Figure 2

Table 4 gives measured amplitude variation and corresponding (C/I) values for locations 3, 4 and 5 at different milling angles.
From these results a "worst case" estimation of the bistatic r.c.s. was derived, using equation (3-4).

location	milling angle	amplitude variation	C/I(dB)	$\sigma(dBm^2)$
3	-30	2.8	8.6	43.9
3	-45	3.1	7.4	43.0
3	-70	4.3	3.9	46.5
3	-80	3.8	5.2	45.2
3	-89	2.9	8.1	42.9
4	-70	1.0	18.3	44.9
5	-70	0.5	24.6	43.0

Table 4. Amplitude variation, corresponding (C/I) and calculated bistatic r.c.s. for locations 3, 4 and 5 at different milling angles.

6. APPLICATION OF THE MEASUREMENT RESULTS

6.1 Introduction

In chapter 5 values of the (bistatic) radar-cross-section (r.c.s.) were derived for the forwardscatter region and the backscatter region.

By writing equation (3-4) in an alternative way:

$$20 \log r = (C/I)_{req} - 10 \log 4\pi + 10 \log \sigma - A_2 + A_1 - A_r + \Delta G \qquad (6\text{-}1)$$

the distance r from the WECS to a location P can be calculated for a required (C/I): $(C/I)_{req}$.
The values of $(C/I)_{req}$ corresponding to the TV picture quality can be derived from table 1.
When different $(C/I)_{req}$ curves around the windturbine are drawn, an impression of the extent of the disturbing effect to UHF-TV reception around the WECS will be obtained.

6.2 Backscatter region

For the backscatter region the maximum value of the r.c.s. of the WECS was estimated at: $\sigma = 24\ dBm^2$.
From (6-1) the distance r follows from:

$$20 \log r = (C/I)_{req} - 11 + 10 \log \sigma - A_2 + A_1 - A_r - \Delta G.$$

For a worst case approximation $\Delta G = 0$ dB (no discrimination of the receiving antenna).
When $A_1 = A_2 = A_r = 0$ dB, r can be calculated from:

$$20 \log r = (C/I)_{req} - 11 + 24 \qquad (6\text{-}2)$$

In the backscatter region the following values of r corresponding to $(C/I)_{req}$ can be derived.

$(C/I)_{req}$[dB]	Δ[dB]	r
> 39	< 0,1	400 m
33	0,2	200 m *
27	0,4	100 m *
20	0,8	50 m *

Table 5

6.3 Forwardscatter region

The bistatic r.c.s. in the forwardscatter region was measured at 7 different locations.
Tables 3 and 4 show that strong variations may occur, dependent on the milling angle of the WECS. When the maximum value of the bistatic r.c.s. is chosen: $\sigma = 46.5$ dBm^2 the $(C/I)_{req}$ contours can be determined from (6-1). For the forwardscatter region $\Delta G = 0$ dB (antenna discrimination is not possible).

When A_1, A_2 and A_r are set to 0 dB, the distance r can be obtained from:

$$20 \log r = (C/I)_{req} - 11 + 46.5 \qquad (6\text{-}3)$$

resulting in table 6:

$(C/I)_{req}$ dB	Δ dB	r
> 39	< 0,1	5,3 km
33	0,2	2,7 km
27	0,4	1,3 km
20	0,8	600 m
10	2,5	200 m

Table 6

6.4 Calculation of v.c.s. from a simple model

In many cases an estimation of v.c.s. of a WECS in both back- and forwardscatter region is desirable without using measurements. For the scattering characteristics of a large WECS an approximation by a metal cylinder seems most appropriate.
For a cylinder the r.c.s. is given by

$$\sigma = 10 \log \frac{2\pi a L^2}{\lambda} \text{ [dBm}^2\text{]}$$

in which a = radius of cross section
L = length of the cylinder
λ = signal wavelength.

It is assumed that interference in the backscatter region occurs mainly by reflection from metal parts in the rotorblades. An estimation of the root construction of the rotorblades of the 45 m WECS by a metal cylinder of length 3.2 m and cross-section radius of .6 m will yield a r.c.s. of:

$$\sigma = 20 \text{ dBm}^2 \text{ when } \lambda = .4 \text{ m.}$$

For the forwardscatter region a representation of the rotorblades by a metal cylinder of length 45 m and radius 1 m leads to a r.c.s. of:

$$\sigma = 45 \text{ dBm}^2 \text{ when } \lambda = .4 \text{ m.}$$

6.5 (C/I) Contours

In figure 2 the (C/I) contours from tables 5 and 6 are drawn on the map showing the area around the 45 m WECS. Some uncertainty still exists concerning the width of the forwardscatter sector; or: 'where does the backscatter region change into the forwardscatter region?'
From measurements at locations 2, 3 and 4 no decision could be obtained due to the effect of a small forest. Observations at location nr. 5 (about 100 m besides the line: transmitter - WECS, at a distance of 875 m from the WECS (line of sight) showed no interference from the WECS.

7. CONCLUSIONS

1. Interference of the 45 m WECS near Medemblik (the Netherlands) on UHF-TV broadcast reception was shown.
2. From measurements on the radar-cross-section (r.c.s.) of the WECS; the (C/I) can be determined in a large area around the WECS. Because the (C/I) is decisive for the picture quality of TV broadcast reception (table 2), contours of constant (C/I) can be drawn around the WECS (figure 2).
3. Measurements revealed that interference from the WECS in an area between transmitter and WECS (backscatter region) is significantly less than in the area behind the WECS, seen from the transmitter (forward-scatter region).
4. In the backscatter region of the 45 m WECS near Medemblik, annoyance can be expected within an area of 100 m (C/I = 27 dB contour). When a minimum (C/I) of 33 dB is taken for acceptable TV reception, an area of 200 m around this type of WECS is recommended to be kept free of viewers. Besides in the backscatter region advantage can be taken from the directive properties of TV antennas to improve the (C/I).
5. In the forwardscatter region interference can be expected from the 45 m WECS at distances up to 1.3 km (C/I = 27 dB contour).
   When the (C/I) = 33 dB criterion is used an area of 2.7 km is recommended to be kept free of viewers. Improvement of TV picture quality in the forwardscatter region by using directional antennas is <u>not</u> possible. Because the (C/I) can vary from place to place at distances up to 1.5 km, improvement of (C/I) can be obtained at some places by shielding of the unwanted (I) signal and/or line of sight path for the wanted (C) signal.
6. It should be noted that for the installation of WECS, clear distinction should be made between the backscatter- and forwardscatter region leading to the "key hole" contour in figure 2. For the backscatter region reflections from well conducting (metal) parts in the rotor blades are important while the size of the forwardscatter region is ruled by the largest dimensions of the rotor blades (diameter).

REFERENCES

(1) "Electromagnetic Interference to Television Reception caused by Horizontal Axis Windmills"
D. Sengupta, T. Senior;
Proceedings of the IEEE Vol. 67, No. 8 Aug. 1979

(2) "Bistatic RCS of complex objects near forward scatter";
J. Glaser;
IEEE Transactions on AES. Vol. AES 21 no. 1
jan. 1985

(3) "Reflections of electromagnetic waves by large windturbines and their impact on UHF-broadcast reception",
P. van Kats, O. de Jager;
Dr. Neher Labs Rept. 511 TM, aug. 1984

(4) "Measurements of impulse response of a wideband radiochannel at 910 MHz from a moving vehicle"
J. van Rees;
Electronic Letters, vol. 22, no. 5, febr. 1986

APPENDIX

Examples of field strength variations due to obstruction of the radio path by the rotorblade rotation of WECS (resp. forwardscatter region figure 3, backscatter region figure 4).

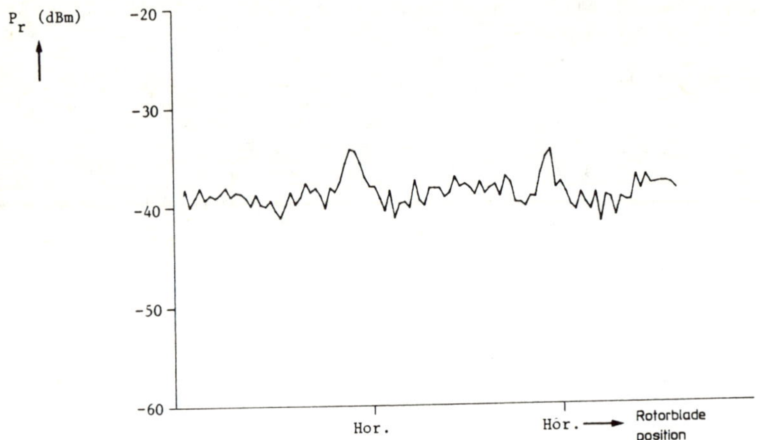

Received power during rotorblade rotation, measurement position nr.3, milling angle 70°, TV1 (620MHz).

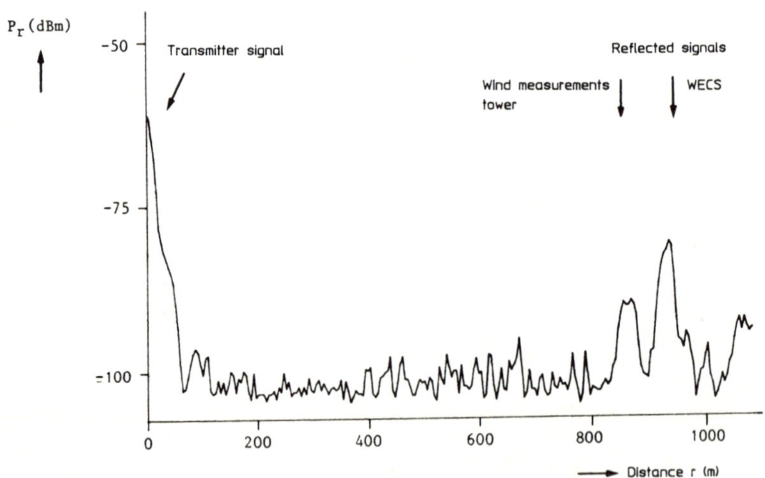

Reflection measurement backscatter region, measurement position nr.8, milling angle 320°, frequency 750MHz

Chapter 12

Recent developments in the Dutch safety regulations for WECs

L. A. H. Machielsc and H. H. Ottens

1. INTRODUCTION

In September 1985 an official draft version of the Dutch safety regulations NEN 6096 for WECS was released (1). It was prepared by NEC 96 a subcommittee of the Netherlands Electrotechnical Committee. The regulations gave (conservative) specifications for the design of WECS by means of a set of requirements concerning electrical systems, safety systems, structural design and labor safety.

After publication of the draft suggestions for improvement were collected for incorporation in a revised version.
Meanwhile a research project called "Design criteria for small WECS" being a part of the National Research and Development Program on WECS (NOW) was finished (2).
This project was concentrated on the generation of simple but reliable descriptions of the atmospheric conditions and the related load calculations for the "design by rule" of WECS.

For the revised version it was decided to change the chapter concerning the requirements on structural safety according to the results of the project "Design criteria for small WECS". Next to it a more fundamental detailed scheme for determination of design load cases was incorporated. In the same time the adopted scheme and the related changes in terminology should be in accordance with the new draft series of building codes.
Therefore chapter 5, dealing with structural safety, has been rewritten. This work has been finished recently and the revised version of NEN 6096 will be published in summer 1988.

This paper gives a brief description of the resulting regulations with special emphasis on the method and philosophy of chapter 5 dealing with structural safety.

2. DESCRIPTION OF THE SAFETY REGULATIONS NEN 6096

The safety regulations NEN 6096 start with a description of subject and scope. The regulations apply to horizontal axis wind turbines in the range of 2 up to 20 m rotor diameter that are grid connected and manufactured in series.
After definitions and nomenclature separate chapters deal with electrical safety, structural safety and labor safety. Another chapter gives measures and recommendations to restrict the

harmful results of lightning strikes. Another chapter was planned on experimental verification of the structural integrity. This part has not been finished yet and will be included in a future edition probably .

Apart from chapter 5 on structural safety there are only small differences compared to the draft. Chapter 5 however was fully changed for two reasons.
First the results of the NOW project "Design Criteria for small WECS" were incorporated. The aim of this project was to define less conservative loads for small WECS compared to the loads given in the draft regulations NEN 6096. As a result characteristic design wind conditions were specified. Also fatigue loads, loads for normal operation during extreme gusts and failure induced loads were calculated. For the failure induced loads distinction was made for different safety systems and different turbine control systems.
The characteristic design wind conditions are incorporated in the regulations, while the calculations of the loads will be published in a separate "guide for practice".
Next to it the methodology and terminology of the new draft safety regulations for buildings (NEN 6700 series) were adapted for definition of load cases and verification of the structural integrity of the turbine by calculation. This was done for the comfort of local authorities that are responsible for licensing and to avoid problems in this matter caused by differences in terminology.

3. THE CHAPTER ON STRUCTURAL SAFETY

Chapter 5 gives the sections dealing with structural safety. The basic philosophy contains the principle that a WECS must be designed in such a way that it can withstand one abnormal load case at a time. Next to it the turbine must remain in a safe condition after a failure of one component or system. In that case the turbine is supposed to fulfill the demands of the Nuisance Act requiring that the surroundings will not incur risks.

4. SAFETY SYSTEM

One part of chapter 5 contains the requirements for the safety and blocking systems of a wind turbine. This part has nearly not been changed compared to the draft version. It states among others that at least two dissimilar safety systems are needed These systems must be not only independent of each other but also be activated independently.

5. DESIGN LOAD CASES

Much attention in the regulations has been payed to the definition of the design load cases in agreement with the draft NEN 6700 series of building codes.

The possible modes of operation of the turbine can be divided into normal modes of operation and particular modes of operation, with a high and an extreme low probability of occurrence respectively. For example normal modes of operation are normal energy production, normal parking and starts and stops. An overview of the particular modes of operation to be considered is given in table 1.

	PARTICULAR MODES OF OPERATION

FAILURE DURING NORMAL OPERATION	pitching mechanism of 1 blade
	pitching mechanism of all blades
	yaw mechanism
	safety mechanism
	turbine control
	blocking of bearings
	frequency convertor
	electrical system
	    short circuit 2 phases
	    short circuit 3 phases
	    disconnection

FAILURE DURING NORMAL PARKING	pitching mechanism of 1 blade pitching mechanism of all blades yaw mechanism safety mechanism turbine control
OTHER	simultaneous action of both safety mechanisms blocked turbine transport and erection

Table 1. Overwiew of the particular modes of operation that have to be considered according to the regulations

The variable external loads also can be divided into loads with a high and an extreme low probability of occurrence. Loads with a high probability of occurrence are fatigue loads and so called momentary loads. Loads with an extreme low probability are called extreme loads.

Six different combinations can be made out of the 2 modes of operation and the 3 groups of loads. Bearing in mind, that the WECS has not to be designed for multiple abnormal conditions, only 4 combinations have to be considered for verification of the turbines structural integrity. These 4 sets of load cases are called momentary, particular, fundamental and fatigue combinations (Table 2). They originate from combinations of possible operational modes and possible external loads acting on the WECS.

MODES OF OPERATION	MOMENTARY LOADS	EXTREME LOADS	FATIGUE LOADS
NORMAL MODES	momentary combination	fundamental combination	fatigue combination
PARTICULAR MODES	particular combination	-------------	-----------

Table 2 The 4 sets of load cases (combinations) that must be considered

6. LOAD SPECIFICATIONS

Three kinds of loads are distinguished in the regulations: loads originated by the wind, by the energy conversion system and other loads.

The regulations present a frequency distribution for the average wind speed with a discrete longitudinal gust model characterized by amplitude and gradient, wind direction changes and vertical wind shear including a stochastic component due to turbulence. This characterization of the wind comes from the project "Design criteria for small WECS". It is the result of a method developed by MT-TNO for characterizing turbulence effects averaged over the rotor area on a statistical basis (3). The method can give deterministic characterizations of turbulence effects with different probabilities of occurrence and therefore has been used for determining design wind conditions for the calculation of the momentary, extreme and fatigue loads. See table 3 with wind data for fatigue calculations.
Next to it the regulation also gives specifications for the effects of tower shadow for up-wind and down-wind turbines, presented as vertical wind shear.

WIND SPEED INTERVAL [m/s]	CHARACTERISTIC WIND SPEED [m/s]	GUST AMPLITUDE [m/s]	GUST PERIOD [s]
up to 8	7	2.3	8
8 - 12	11	3.5	6
above 12	15	4.7	4

also to be considered the effects of: tower shadow
wind direction shift
vertical wind shear

Table 3 Longitudinal gust data for the calculation of the fatigue loads

All the characteristic design wind data in the code are determined for a reference turbine of 16 m diameter and hub height of 20 m. The frequency distribution for the average wind speed of Den Helder has been adopted as a basis for design calculations, being the location with the highest average annual wind speed. The calculation of turbulence effects in most cases is based on the assumption of a surface roughness length of 1 m because of the possibility of obstacles in the surroundings.

Two loads induced by the energy conversion system are considered in the regulations, namely the loads induced by a disconnection of the grid and loads caused by short circuit in the generator. The first situation is considered to give momentary loads, so the probability of occurrence is not thought to be extremely low. This indeed can be the case when the turbine is connected to a weak grid.
A short circuit has a very low probability of occurrence and is therefore assigned to the extreme loads.

Table 4 gives a summary of the characteristic effects caused by the wind and the energy conversion system that must be taken into account.

The standard also gives specifications for other kinds of loads originating from atmospheric, environmental conditions, hail, snow and ice.

	VARIABLE LOADS		
ORIGIN OF LOADS	MOMENTARY	EXTREME	FATIGUE
average wind speed	*	o	*
amplitude gust	*	o	*
gradient gust		o	
vertical wind shear	*	o	*
wind direction change	*	o	*
tower shadow	*		*
loss of grid	*		
short circuit		o	

* to be combined with other loads of the same kind
o to be combined with momentary loads of different origin

Table 4 Variable loads originating from wind conditions and energy conversion system

7. STRUCTURAL INTEGRITY CHECK

The structural integrity of the turbine must be demonstrated by calculation. For that purpose the general formulation of the draft NEN 6700 series is adopted:

$$S (Fd , GEd , Md) < R (Fd , GEd , Md)$$

in which S and R are the load and strength function respectively and Fd, GEd and Md calculation values for the loads, the geometrical parameters and the material properties.
The regulations require that limit states for operation shall not be exceeded for the momentary combinations, and failure limit states shall not be exceeded for the particular, fundamental and fatigue combinations.
Limit states for operation are the limitations for normal use of the turbine. Failure limit states are limitations bounded by failure of components or structural elements.

The safety margins in the calculation are explicitly determined by the use of material and load factors.
The load factors have different values for operational and failure limit states. Also a distinction has been made in load factors that must be applied in the calculations for tower and foundation (table 5). If the probability of injuries and economical damage is low the designer can use decreased load factors in the calculations for these turbine parts. The remaining parts must be designed for safety class 2.
Material factors also are specified for the 2 different limit states to be considered (table 6).

LOAD FACTORS

LOADS	FAILURE LIMIT STATES		OPERATION LIMIT STATES
	safety class 1	safety class 2	all classes
momentary loads	1.2	1.3	1
extreme loads	1.2	1.3	1
fatigue loads	1.2	1.2	1

Table 5 Load factors to be used in the calculation of the variable loads

MATERIAL FACTORS

MATERIAL	FAILURE LIMIT STATES	OPERATION LIMIT STATES
steel	1.0	1.0
concrete (reinforcement)	1.1	1.0
concrete (compression)	1.1	1.4
laminated wood	1.2	1.2
aluminum	1.0	1.0
reinforced plastics	1.2	1.2

Table 6 Material factors to be used in the calculation of the material properties

8. FURTHER DEVELOPMENTS

Recently a more detailed description of the design wind data has been published (4). The guide presents the data in nomograms. It gives a designer the opportunity to determine the wind data depending on the actual rotor diameter up to 25 m, hub height up to 40 m and roughness length of the surroundings. Next to it a distinction has been made between coastal and land locations. The data are expected to be very useful in combination with time domain simulations of the behavior of the turbine.

A this moment a program is started to verify the blade fatigue loads calculated from the wind data given in the regulations together with the load calculations given in the guide of practice mentioned in paragraph 2. Also the design wind data from (4) in combination with time domain simulations will be subjected to a verification. A comparison will be made with measured results of a 8.7 m and a 25 m diameter wind turbine.

9. CONCLUSIONS

The new safety regulations NEN 6096 offer the designer the opportunity of a less conservative and more detailed determination of design loads, especially for fatigue calculations. The definition of safety margins has an explicit form using material and load factors.
The method and terminology are in agreement with coming codes for buildings that will become widely known among local authorities.

10. REFERENCES

1. Safety requirements for wind turbines,
   Netherlands National Committee of IEC and CENELEC
   concept standard NEN 6096, 1985 (in Dutch)

2. Design criteria for small wind turbines,
   Part 1: Overview,
   H.H. van der Linden,
   NLR TR 86112 U (in Dutch)

3. Design criteria for small wind turbines,
   Part 2: Wind conditions,
   G.L.H. Beugeling, P.E.J. Vermeulen,
   MT-TNO Report 85-010880, 1985 (in Dutch)

4. draft: Handbook design data for wind turbines,
   part 1,
   A.E. Pfeiffer et al,
   Holland Windturbine BV, MT-TNO,
   November 1987, (in Dutch)

Chapter 13

An analysis of blade throw from wind turbines

D. M. Turner

SYNOPSIS

The CEGB is investigating the potential of large wind turbines for generating electricity. The most serious failure from the safety point of view is the detachment of a blade or blade fragment which could be thrown a considerable distance and could damage people or property. This is however an extremely unlikely event though it has been given considerable attention over several years. Though a few wind turbine blades have become detached in the past there is a current trend of fewer incidents recently as operating experience is gained and wind turbine designs mature. This paper first reviews earlier work on wind turbine safety. The aerodynamical calculations are described and some consideration of bouncing and skidding upon impact is given. A six degrees of freedom model for following the tumbling motion of the blade fragment is described. The model is better than ones previously described in that the angular velocity of the fragment is included in the calculation of the angle of attack of the wind.

Blade fragment maximum throw distances are typically less than 1 km which is less than 25% more than the calculations which neglected the effect of the lift force. In all cases a stable rotation was observed about either the minor or major axis of inertia. Even starting with zero initial angular velocity, rotation about the minor axis of inertia was developed. Better aerodynamical data is required for rotating fragments to increase confidence in the predictions of the six degrees of freedom model. Experimental data for comparison would be very useful.

1. INTRODUCTION

The CEGB is investigating the potential of large wind turbines for generating electricity (Milborrow, 1982). The choices of machine design and site location will be made with full consideration of the safety implications for the public and CEGB operating staff. The most serious failure from the safety point of view is the detachment of a blade or blade fragment which could be thrown a considerable distance and could damage people or property. This is however an extremely unlikely event though it has been given considerable attention over several years. Macqueen Ainslie, Milborrow, Turner and Swift-Hook (1983) have provided a comprehensive review of the risks associated with wind-turbine blade failures. This includes a review of operating experience. Though a few wind turbine blades have become detached in the past there is a current trend of fewer incidents recently as operating experience is gained and wind turbine designs mature.

The purpose of this paper is two fold. Firstly the previous work of Macqueen et al. (1983) and Turner (1987) will be reviewed. Secondly a full six degrees of freedom trajectory model is described which follows the tumbling motion of the blade fragment. In earlier work the aerodynamical calculations used averaged drag parameters for calculating the effects of aerodynamic force on the blade fragment. Previously it was argued from energy considerations, Macqueen et al. (1983), that the blade fragments would tumble along their trajectory. This is indeed seen to be the case with the lift force increasing the throw distances by less than 25% as stated by Macqueen et al. (1983).

2. REVIEW OF PREVIOUS WORK

The probabilities calculated in this paper show the risk to a particular area given that the blade fragment has broken from the rotor. Thus to calculate the actual risk to a given area these probabilities must be multiplied by the probability of the fragment first becoming detached. Macqueen et al. (1983) estimated that the probability of a blade becoming detached was less than 10^{-5} per machine per year. As indicated in the introduction it is believed a lower figure is now applicable. The worst case failure occurs when the wind turbine accidentally becomes disconnected from the grid and a blade fragment breaks from the overspeeding rotor. At a tip speed of Mach 0.82 the aerodynamic torque on the blade is zero. This provides a limiting tip speed for runaway rotors and thus an upper bound to the throw velocity for fragments. Indeed it is likely that a well designed rotor will break due to the centrifugal force before this limiting tip speed is reached, thus reducing the throw velocity of the fragment. It should be remembered that for a horizontal axis wind turbine there is only a relatively few release angles which can give rise to a significant throw distance. In most cases the fragment will be thrown to the ground in the vicinity of the wind turbine tower. This is not the case for vertical axis wind turbines but these benefit since there is no possibility of a large throw distance due to a fragment being thrown upwards at about 45 degrees.

The motion of a fragment moving under the influences of gravity and drag with velocity $\underline{V}$ in a wind of velocity $\underline{W}$ is given by

$$d_t\underline{x} = \underline{V}$$

$$d_t\underline{V} = \underline{g} - \mu\lfloor\underline{V} - \underline{W}\rfloor(\underline{V} - \underline{W}) \qquad \ldots (2.1)$$

where $\underline{g} = (0, 0, -g)$, g is the acceleration due to gravity, μ = drag parameter = $\frac{\frac{1}{2}C_D\rho_a A}{M}$, C_D the drag coefficient, ρ_a the air density, M the fragment mass and A the plan area of the fragment. Turner (1987) describes a Monte Carlo method for assessing the risk posed by failure. The probability distribution is built up from many solutions of equations (2.1) under a variety of conditions. Before each solution of equation (2.1) there are five quantities which must be specified

1. Wind Direction
2. Wind Speed
3. Breaking Point along the blade
4. Angle of rotation at which the break occurs
5. Angular velocity of the blade.

The safety assessment is based upon a consideration of both the results from the Monte Carlo simulation and a consideration of particular worst cases, e.g. the maximum throw distance for a fragment released at the maximum speed with the once-in-fifty years gust of 40 ms^{-1}. It is no use putting extremely low probability events into a Monte Carlo calculation as they are not likely to be included in any finite number of simulations.

The probability of any particular wind direction and speed can be obtained from the meteorological data for a particular site and the wind speed can be correlated with the wind direction. The radial distance along the blade at which the break is assumed to occur should be obtained from a probability distribution derived after an assessment of the blade design. Bolted joints or welds are weaknesses in the blade construction and the places where they occur are potential breaking points. Historical evidence concerning blade failure shows that the breaks occur mainly close to the root. In the absence of detailed data a geometrical distribution is assumed starting at the tip. It is then assumed that the blade is twice as likely to break at the next joint inwards etc. up to the hub. Macqueen et al. (1983) calculate the value of the drag parameter μ for both whole blades and tip fragments. Note that the drag parameter for the tip fragment is always larger than for the whole blade. Thus the difference in the maximum throw distance for the whole blade and tip fragment is not as large as may be expected.

A uniform probability distribution is used at present to describe the likelihood of the blade breaking at various angles of rotation. If the wind is set to zero and both the breaking point on the blade and the angular velocity of the blade are fixed then the probability distribution for impact at a given range can be obtained purely as a function of the release angle. A particular case is shown in Fig. 1. It can be seen that the probability is relatively high near the tower, low in the middle of the range and increases towards the maximum range.

These are two main factors which contribute to the assumed probability distribution for the rotation rate of the rotor at the moment of failure, these are

(i) The likelihood of the blade rotating at a given angular velocity.

(ii) The likelihood of a break occurring at a given angular velocity.

The angular velocity of the blade remains constant at the rated speed whilst the turbine is in operation. Only for a very small fraction of the time will the wind turbine rotate at speeds in excess of the rated maximum before feathering of the blades or the braking system reduces the rotation rate. The second of the above factors accounts for any substantial changes in the general stress levels due to increased rotation rate or larger torques at higher wind speeds. Thus a failure is more likely to occur at an overspeed or in a high wind though these conditions only exist for a small period of time. Further details are given by Turner (1987).

Fig. 2 shows typical results obtained from the Monte Carlo calculation together with calculated confidence intervals for each Sector. Note that in the simulation no fragment has gone further than 700 m. For the supplied input data the worst case would have resulted in a fragment been thrown over 700 m. This illustrates the earlier point that the larger throw distances have an extremely low probability of occurrence and should be considered individually. In fact the data used to obtain the probabilities in Fig. 2 are believed to be slightly conservative in the calculation of the likelihood of high release speeds.

The Monte Carlo calculation may also perform some assessment of the bouncing and skidding of the fragment upon impact. It is difficult to make precise predictions about the behaviour of the fragment upon impact. However it seems reasonable to describe two modes of motion, (1) bouncing and (2) skidding after impact.

A cannon ball is the closest approximation to a bouncing fragment for which data can be found. Assessment of mid 18th-19th Century data on gunnery trials, obtained by kind permission of the Royal Artillery Library, Woolwich, indicates that:

(i) There is a cut-off angle of impact above which the cannon ball will not bounce, slightly less than 20°;

(ii) The angle with which a cannon ball leaves the ground after bouncing is double the angle of impact on approach. It can be seen that because of the effect of drag and ignoring the effect of lift, at each bounce the angle of impact is at least double the angle for the previous bounce so that the cut-off angle is reached after a few bounces.

The cannon ball data applies to 'typical' field conditions which should be no firmer than the 'agricultural' land likely to surround a wind turbine. The proportion of tarmacadam area round a turbine is expected to be very small and does not require separate treatment.

Based on this approach it is assumed that the fragment will bounce if the angle of impact is less than a suitable cut-off angle, usually 20°.

When the angle of impact of the fragment with the ground is greater than the cut-off angle the fragment will slide. This happens whether or not bouncing has occurred previously. The sliding distance s of the fragment is evaluated by applying a constant retardation, -a, to the tangential velocity U_t upon impact and thus $s = U_t^2/2a$. The acceleration, a, is supplied by the user; typically g/2 is used, Eggwertz et al (1981).

The difference in height of various objects in the vicinity of the turbine may also be included.

The effect of the object height is calculated by considering the distance in front of the first impact at which the fragment might possibly hit an object or person of height h_o. For a fragment where the largest diameter is l this is assumed to occur at a height $d_o = h_o + l/2$ above the ground.

A danger rectangle is calculated. The length of the rectangle is d_o plus the distance bounced plus the sliding distance. The width is taken to be the largest dimension of the fragment. However if the fragment is small this width is set to a nominal minimum value. Assuming that the danger rectangle intersects a particular segment of the output, then the probability of hitting an object in the segment is obtained from a linear interpolation. An area interpolation is made between the probability of hitting a point object in the segment and the certainty of hitting an object occupying the whole segment. Further details are given by Turner (1987). Typical results obtained from the Monte Carlo simulation including bouncing etc. are given in Fig. 3 and show that even if a blade fragment is detached the probability of a person in the vicinity being hit varies from 10^{-2} close to the tower to 10^{-5} at 600 m. It should be remembered that these figures are averages. Obviously if a person is standing close to the tower but upwind of the turbine then the probability of being hit is very low. However if someone is close to the tower and just downwind of the rotor plane when a fragment detaches they are more likely to be hit. Obviously at sites with a strong prevailing wind a horizontal axis turbine is likely to be in a particular orientation and thus some areas will be much less likely than others to be hit by any detached blade fragments.

In this Section a probabilistic approach has been described. The remainder of this paper is concerned with a 3 dimensional aerodynamical calculation of a tumbling blade fragment to describe an extremely low probability worst case situation.

3. THREE DIMENSIONAL AERODYNAMIC CALCULATION

This Section describes a calculation which can actually predict and follow the tumbling motion of a detached wind turbine blade fragment along its trajectory.

Aerodynamical calculations of trajectories have other applications, e.g. boomerangs, Hess (1968) and tornados, Redman et al. (1978). The angular rotation of the boomerang remains relatively constant during the motion and thus Hess 1968 could describe the path without evaluating the angular velocity at each time step. This provides a simpler model for the motion than the one discussed here. Redman et al. (1978) use a six-degrees of freedom model to describe the transport of objects by tornados. They obtained three-dimensional aerodynamic data for three objects, a metal cylinder, a plank of wood and a car. The data collected is not directly applicable to the wind turbine problem. It should be noted that Redman et al. (1978) calculate the angle of attack using only the transational velocity of the centre of gravity of the object. Thus it is assumed that the objects momentarily have no angular velocity and a separate 'damping coefficient' is supplied for cases of rapid rotation. By contrast here the angular velocities are used in the determination of the angle of attack. Sorenson (1984) describes a six degrees of freedom model for wind turbine blade trajectories. He also uses aerodynamic coefficients averaged over the whole of the blade fragment. Such a fragment may be 50 m long and when rotating the wind velocity could be markedly different at either end. Thus a single averaged angle of attack velocity may not be appropriate and here a model permitting some subdivision of the blade fragment span is described.

3.1 Rotating Frames of Reference

To perform 3-dimensional aerodynamic calculations of blade fragment motion it is convenient to consider two coordinate systems. These coordinate systems are shown in Fig. 4. They will be referred to as inertial (x, y, z) and body (1, 2, 3) coordinates. The direction of the latter will be denoted by $(\underline{i}, \underline{j}, \underline{k})$. The origin of the body coordinates is taken to be the centre of gravity of the blade fragment, these axes remain fixed in the body as the blade fragment tumbles or spins along its trajectory.

The equations for linear motion are most easily written down in the inertial coordinates. Newtons second law of motion gives simply

$$\underline{F} = m \frac{d\underline{V}_s}{dt} \qquad \ldots (3.1)$$

where $\underline{F}$ denotes the external forces acting upon the blade fragment and $\underline{V_s}$ its velocity in the inertial (or space) coordinates. The $(\underline{i}, j, \underline{k})$ coordinate system will be assumed to be fixed at the instantaneous position of the centre of gravity of the blade fragment. Suppose the velocity vector $\underline{V_s}$ of the blade fragment at some instant is given by

$$\underline{V_D} = V_1\underline{i} + V_2\underline{j} + V_3\underline{k} \qquad \ldots (3.2)$$

Synge and Griffith (1959) provide the following relation between inertial and body coordinates

$$\left[\frac{d}{dt}\right]_{space} \underline{V}_s = \left[\frac{d}{dt}\right]_{body} \underline{V}_s + \underline{\omega} \wedge \underline{V}_s \qquad \ldots (3.3)$$

Using (3.2) and (3.3) equations (3.1) become

$$\begin{aligned} m(\dot{V}_1 - V_2\omega_3 + V_3\omega_2) &= F_1 \\ m(\dot{V}_2 - V_3\omega_1 + V_1\omega_3) &= F_2 \\ m(\dot{V}_3 - V_1\omega_2 + V_2\omega_1) &= F_3 \end{aligned} \qquad \ldots (3.4)$$

where the forces F_i, i = 1,.,3 include components from both gravity and aerodynamics.

If the body axes are taken to be the Principle Axes of Inertia then Eulers equations for rigid body rotation are

$$\begin{aligned} A\dot{\omega}_1 + (C-B)\omega_2\omega_3 &= G_1 \\ B\dot{\omega}_2 + (A-C)\omega_1\omega_3 &= G_2 \\ C\dot{\omega}_3 + (B-A)\omega_1\omega_2 &= G_3 \end{aligned} \qquad \ldots (3.5)$$

where A, B, C are the Principle Moments of Inertia about the three axes $\underline{i}$, $\underline{j}$, $\underline{k}$ respectively. Once the Forces F_i and Moments G_i for i = 1,.,3 have been calculated then the six equations (3.4) and (3.5) may be solved for the three linear velocities V_i, i = 1,.,3 and three angular velocities ω_i, i = 1,.,3. The forces and Moments are calculated at every time step during the trajectory and thus the tumbling motion of the blade is followed.

Suppose at time t the rotation is given by the motion $R^{(n)}$. The rotation matrix $R^{(n+1)}$ at time t + Δt is evaluated from $R^{(n)}$ by

$$R^{(n+1)} = \Phi_3\Phi_2\Phi_1 R^{(n)} \qquad \ldots (3.6)$$

where each Φ_i, i = 1,.,3 is a 3x3 matrix representing a rotation about the ith body axis. Equation (3.6) is an approximation valid for small angles, thus small values of $\omega_i\Delta t$, i = 1,.,3. After each update of equation (3.6) the new rotation matrix $R^{(n+1)}$ is renormalised otherwise the accumulation of rounding errors cause the determinant to depart from unity. The gravitational force is transformed to the body coordinates using the matrix $R^{(n)}$. As indicated the gravitational force is included on the right hand side of equations (3.4). The solution of equations (3.4) and (3.5) yields the linear velocity relative to the instantaneous body axes and the matrix $R^{(n)}$ is used to transform to the inertial coordinates and thus locate the blade fragment relative to the wind turbine tower.

3.2 Aerodynamic Data

The blade fragment is considered as two panels as shown in the plan form of Fig. 4. For convenience the smooth curved shape of the panel has been approximated by two rectangles. Also the two panels are chosen so that the centre of gravity of the blade fragment lies on the line joining the two panels. The main reason for considering the two panels is that it takes some account of the changing incident wind speed along the blade fragment span. Thus when the blade fragment is spinning rapidly around axis 3 (as is initially assumed) the incident wind speeds on the two panels can be significantly different.

In Section 3.3 the equations for the case shown in Fig. 4 are derived. Here a force is calculated at the centre of each panel and then a resultant and moments evaluated. The panels are assumed to lie so that the line joining their individual centres of gravity is parallel to the front and rear edges of the rectangles.

Ideally for solving this problem full three dimensional aerodynamic data at different Reynolds numbers should be available for each part of the blade at different rotation rates around the body axes. Such comprehensive data are seldom (if ever) available; typically only steady state data for a fixed Reynolds number for the lift, drag and quarter chord moment coefficients are published. This would appear to be a minimum amount of data for a calculation including lift. The evaluation of the aerodynamic forces acting on the blade is described below.

3.3 Aerodynamic Forces and Moments

To obtain a solution using the above minimum data the three dimensional flow is approximated by a pair of two dimensional flows. First flow in the (2,3) plane for each panel in Fig. 4 is considered. This is the usual two-dimensional flow over an aerofoil considered in text books (e.g. Milne-Thompson 1966). Secondly a drag force is assumed to act along the blade fragment span which will be described in Section 3.4.

The modulus of each aerodynamic force $\underline{F}$ acting over a panel of plan area A_F is given by an equation of the form

$$| \underline{F} | = \frac{1}{2} \rho A_F C_F | \underline{V} |^2 \qquad \ldots (3.7)$$

where ρ is the air density, C_F the force coefficient and $\underline{V}$ is the incident wind velocity relative to the centre of mass. In this model it is assumed that forces in the (2,3) plane may be calculated using

$$V = \sqrt{V_2^2 + V_3^2}\ .$$

In 2-dimensional aerodynamics it is usual to consider lift (L) and drag (D) forces. These forces are evaluated from the force coefficient data for C_L and C_D. These coefficients have been tabulated at 5° intervals which are linearly interpolated to produce coefficients for any angle of attack. Fig. 5 shows the eight different cases for a two dimensional aerofoil. The directions in which the lift and drag forces act are shown for a given incident wind velocity to a stationary blade. Using the trajectory and aerofoil dimensions the force coefficients are used to calculate forces acting along the body centred axes the forces are evaluated at each time step during the calculation. The lift and drag coefficients may be combined to produce force coefficients C_2, C_3 for the forces acting along axes 2,3 respectively. These force coefficients are given by

$$\begin{aligned} C_3 &= C_L \cos\alpha_{23} + C_D \sin\alpha_{23} \\ C_2 &= C_D \cos\alpha_{23} - C_L \sin\alpha_{23} \end{aligned} \qquad \ldots (3.8)$$

where α_{23} is the angle of attack in the (2,3) plane, thus

$$\begin{aligned} S_1^2 &= (V_3 - \tfrac{1}{2} L_1\omega_2)^2 + (V_2 + \tfrac{1}{2} L_1\omega_3)^2 \\ S_2^2 &= (V_3 + \tfrac{1}{2} L_2\,\omega_2)^2 + (V_2 - \tfrac{1}{2} L_2\omega_3)^2 \end{aligned} \qquad \ldots (3.9)$$

$$\sin\alpha_{23} = \frac{(V_3 + (-1)^i \frac{1}{2} L_i\omega_2)}{S_i} \qquad \cos\alpha_{23} = \frac{(V_2 - (-1)^i \frac{1}{2} L_i\omega_3)}{S_i}$$

for panels i = 1,2. Here ω_i is the angular velocity about the ith body axis and L_i is the span of the ith panel.

Substituting (3.8) and (3.9) into (3.7) gives the following expressions for the forces $F_2^{(i)}, F_3^{(i)}$, i=1,2 along lines parallel to the axes 2,3 respectively

$$F_2^{(i)} = \frac{1}{2}\rho A_i S_i \left[C_D(V_2 - (-1)^i \frac{1}{2} L_i\omega_3) - C_L(V_3 + (-1)^i \frac{1}{2} L_i\omega_2)\right]$$

$$F_3^{(i)} = \frac{1}{2}\rho A_i S_i \left[C_L(V_2 - (-1)^i \frac{1}{2} L_i\omega_3) + C_D(V_3 + (-1)^i \frac{1}{2} L_i\omega_2)\right]$$

for blades $i = 1,2$ ... (3.10)

The data used to evaluate the aerodynamic forces have been obtained from the NACA 0012 aerofoil. These data (Sheldahl and Klimas 1981) are available for a full 180° variation in the angle of attack. This is sufficient for a symmetrical aerofoil.

In the present model a 'drag' force is assumed to act along the panel span. This force does not provide any contribution normal to the panel and is thus not directly opposing the incident wind velocity. All the contribution to the force normal to the blade is assumed to be obtained from consideration of the (2,3) plane. Thus in the absence of force coefficients for the span-wise direction, the aerodynamic force along the span is assumed to be

$$F_1 = \frac{1}{2}\rho A_1 \tilde{V} V_1 C_{D1} \qquad \tilde{V} = \sqrt{V_1^2 + V_3^2}$$

C_{D1} = constant drag coefficient.

Since the wind velocity $\underline{W}_S$ in the inertial coordinates is constant then $[\frac{d}{dt}]^{\underline{W}}s = 0$ and the velocity V_i in equations (3.4) may be replaced by $V_i = W_i - V_i$. Thus equations (3.4) are solved for the velocity of the wind relative to the centre of mass of the fragment. This is precisely the quantity required for evaluating the aerodynamic forces as can be seen from the above discussion.

The moment M of the aerodynamic forces acting on the blade is determined from the quarter chord moment coefficient C_{MQ}, M is given by

$$M = \frac{1}{2}\rho A_F V^2 c\, C_{MQ} \qquad \text{... (3.11)}$$

ρ, A_F, V are defined in equation (3.7). c is the chord of the aerofoil. The chord-wise distance Y of the centre of pressure from the leading edge is given by

$$Y/c = 0.25 - C_{MQ}/c \quad \dots (3.12)$$

The moments of the aerodynamic forces about the body centred axes are evaluated by interpolation of tabulated data of the quarter chord moment coefficients, analogous to the above force calculation. The total moment about the 1-axis is obtained by adding the moments for each part of the blade. The moment about the 2-axis is obtained from the aerodynamic forces normal to the panel. These forces are assumed to act half way along each panel as shown in Fig. 4; thus the moments about axis 2 can be obtained knowing the blade fragment dimensions. Similarly the moments about the 3-axis are calculated from the forces in the plane of the blade. In the present model the force acting along the span is assumed to act through the origin of the body coordinates thus no moment is calculated. As better aerodynamic data becomes available it is expected that the above force calculation can be improved to use the better data.

A numerical solution of equations (3.4) and (3.5) can now be calculated stably with a computer to predict the motion of the blade fragment.

3.5 Qualitative Description of Solutions

The solutions obtained from the equations described in Section (3.2) are dominated by the gyroscopic forces. This can be seen, not only by performing runs excluding the gyroscopic terms, but also looking at the magnitude of the entries on the left hand side of equation (3.5). Unfortunately this does not imply that only the gyroscopic terms are important in the six equations. To obtain a qualitative idea of the solutions to be expected from the six equation model it is necessary to understand the various simpler models which have been put together. For this reason a discussion of the force free Euler equations and rigid body dynamics is given in Appendix A. It is well known that in this case the body rotates steadily about either the minor or major axis of inertia. Precisely which of the two axes the body rotates around is governed by an inequality between constants of the motion. When forces are present the same inequality exists, but the terms of the inequality are no longer constant. Thus if the cumulative effect of the forces over several time steps changes the sign of the inequality then the calculated motion of the blade changes significantly. The blade undergoes a transition and starts to spin about another axis. Thus both the forces and the gyroscopic terms are important.

The motion is affected by both the magnitude and relative magnitude of the moments of inertia. These should be intermediate between the two extremes of a cylinder and a thin flat plate. In the first case the moments of inertia B,C about the body axes 2 and 3 are equal and there is no unique largest axis of inertia. In the second case using the perpendicular axis theorem the moments of inertia satisfy $C=B+A$. This provides the variation in relative magnitude to be expected from the moments of inertia which thus satisfy $0 < C - B < A$ in the notation used here.

3.5 Quantitative Description of the Results

The data for the NACA-0012 aerofoil (Sheldahl and Klimas, 1981) which is typical of realistic wind turbine blade designs, has been used as a test case for the program. To obtain reasonable results the blade mass and moments of inertia given to the program should be consistent with each other and the dimensions specified for the blade. The mass and dimensions of various blade sections are easily obtained from design information together with the moment of inertia of the whole rotor during operation. Using this information the moments of inertia (see Table 1) for the thrown blade may be estimated. Since the moments of inertia B,C are nearly equal it is difficult to estimate the magnitude of the A-inertia term accurately. Thus a number of sensitivity studies of the throw range to the moments of inertia have been performed. These are described later in this Section.

The initial throw orientation, $R^{(o)}$, tower height, wind velocity, blade diameter and release velocity remained the same for all the tests with the values given in Table 1. These values correspond to a failure under very severe conditions when the turbine becomes disconnected from the grid during operation.

The rotation matrix relating the two sets of axes is only first order accurate. Thus high accuracy of the results cannot be obtained simply by using a high order integration scheme for solving the differential equations. A first order implicit scheme and a trapezoidal scheme have both been used for solving equations (3.4) and (3.5). Since a small time step is used the results agree well and there is no obvious sign that the first order scheme is introducing more numerical damping than the trapezoidal rule. However the results discussed here have been obtained using the trapezoidal rule.

The first case considered had an initial angular velocity. $\omega_3 = -3$ rad/s and it can be seen from Table 2 that a time step of 1 millisecond is sufficient to produce a consistent solution. The range changes very little for the three cases considered but the rotation rates about axis 1 vary considerably. In most cases evaluated the blade was spinning about axis 1 on impact. The change to rotation about this axis occurs usually about 0.5 s after the release. The direction of rotation is determined by the direction in which the blade was oscillating about axis 1 at the time the inequality (A1) changes sign. The rotation rate maintains the same sign as when the transition occurs.

The second case considered had an initial angular velocity $\omega_3 = -10$ rad/s. This corresponds to the expected angular velocity if the blade is released with no energy loss at the limiting tip speed of about Mach 1. In this case for the results in Table 2 with a time step of 5×10^{-3} seconds the blade rotated about axis 3 for the whole flight with the inequality remaining positive. However at smaller time steps the blade was calculated to rotate about axis 1 after about 0.5 S of flight with the inequality (A1) changing sign at this point. The changing sign of inequality (A1) effectively produces a bifurcation of the solution since the sign of the inequality dramatically alters the subsequent motion. The accurate reproduction of experimental data for rapidly spinning blades will only be successful if the sign of the inequality (A1) can be controlled properly. However despite the motion about either axis of rotation the range of the fragment is not dramatically increased.

The evidence to date suggests that soon after release a blade will start to rotate about its pitching axis, 1. This adds support to the theory used to justify the particle drag models (Macqueen et al., 1983, Proctor and Turner, 1984). However it seems possible that a blade could remain rotating around axis 3 thrown in 'knife throwing' mode. All the results obtained showed a stable rotation around either axis 1 or 3. The blade never achieved a stable flight with no angular rotation, so extreme ranges are not realized.

As previously indicated a number of sensitivity studies on the Moments of Inertia have been performed. In the preceding examples it has been assumed that C = B+0.5A appoximates the relation between the moments of inertia B and C as discussed in the previous Section. Results for two other cases with C=B+0.05A and C=B+0.95A are presented in Table 2. Thus the effect of the relation between B and C on the flight is obtained from cases 5,7,8. It can be seen that these changes have a significant effect on the range the fragment is thrown, decreasing or increasing the range by about 100 m. Also the calculated rotation rates about axis 1 are quite different. The sign of the rotation rates is the same as that of the oscillation when the inequality (A1) changes sign. At present there is no simple explanation for the effect of the moments of inertia upon the magnitude of the rotation rate around axis 1.

The sensitivity of the motion to the moments of inertia was further investigated by changing the magnitude of A,B by ±20%. Considering the moment of inertia A first, the results are provided in cases 5,9,10 of Table 2. The total range changes by ±5% in Table 2, which is relatively small; however the calculated impact position (x,y) is more substantially affected in Table 2, to the extend -30% to +15%. As may be expected the motion is fairly sensitive to this moment of inertia with the fragment rotating about axis 1 for most of the flight time for the results in Table 2. Comparing cases 5,11,12 shows the effect of changing the moment of inertia B. These results suggest that the actual magnitude of the moment of inertia A and the relative magnitudes of B and C are the two most important features of the moments of inertia which affect the subsequent motion of the blade fragment.

All the results obtained thus far have used a side drag parameter $\mu=0.001\ m^{-1}$. This parameter has been used in the present model since no three-dimensional aerodynamic data are available. It is smaller than the parameter used in the particle drag calculations. This parameter was deliberately underestimated to reduce the effect on the calculation of the drag force along the blade span. In so doing it was hoped that this parameter would then not dominate the calculation.

The results obtained have been compared to a particle with drag calculation. The predicted throw distance with a drag parameter $\mu=0.003\ m^{-1}$ is 630 m and with $\mu=0.002\ m^{-1}$ is 790 m. Using the data in Table 1 these ranges are reproduced here with the blade fragment given no initial angular velocity. On impact the blade has achieved a slow steady rotation about axis 1. Thus when the angular velocity is low the side drag parameter used is having a significant effect upon the range. The forces calculated from the other directions are averaging close to the particle model and the assumptions for the validity of the particle model are being justified. Using the side drag parameter $\mu=0.002\ m^{-1}$ results for initial angular velocities ω_3=-3 and -10 rad/s are presented in Table 2. A time step of 1×10^{-3} s, has been used in both cases which previously yielded results of reasonable accuracy.

In general for the results in Table 2 the higher rotation rates tended to decrease the throw range. When the blade is spinning faster the quadratic nature of the aerodynamic forces yields larger drag forces on the blade leading to a reduced throw range. When the blade is rotating the lift force averages to zero with the present steady state aerodynamic data. An increased range will be obtained from aerodynamic data for which the lift force does not average to zero over a cycle, see Iverson (1979). The reduced throw range is likely to be an aerodynamic rather than numerical effect. It is suggested that rotation dependent aerodynamic data will be needed to accurately reproduce experiments of rotating blade fragments and comparison with experiments seems desirable. Even though drag parameters have been underestimated in these calculations the throw ranges have been increased by less than 25% of the previously calculated values.

4. CONCLUSIONS

1. Previous work on wind turbine blade throw has been briefly reviewed. It is noted that even accounting for possible bouncing and skidding upon impact and given that a failure has occurred then the probability of a person being hit varies from 10^{-2} near the tower to 10^{-5} at 600 m. Assuming a probability of a failure of 10^{-5} these figures reduce to 10^{-7} and 10^{-10} respectively.

2. A six degree of freedom model has been developed for blade fragment motion which includes the lift force, gyroscopic forces and an angle of attack dependent upon both the rotation rate and the translational velocity. The model here is better than those previously described since a realistic calculation of the aerodynamic moments acting upon the blade is performed.

However better aerodynamical data is required for rotating fragments to increase confidence in the predictions of the six degrees of freedom model. Experimental data for comparison would be very useful.

3. The qualitative behaviour of the solutions has been inferred by considering the force free Euler equations. When forces are present there may be a bifurcation of the solution which explains why small changes in the input data may sometimes cause large changes in the flight. The range is less sensitive to the mode of flight.

4. At low initial rotation rates the throw ranges obtained here are close to those obtained with particle drag model calculations.

5. In all cases a stable rotation about either the minor or major axes of inertia has not been observed. Even starting from zero initial angular velocity, rotation about axes 1 was developed.

6. The drag parameter used for the component of flow along the blade span was deliberately under-estimated in these calculations. This was done so that the calculated range was determined from the lift and drag forces acting along the blade chord and not the drag force assumed along the blade span. Even with this under-estimate the throw distances calculated are less than 1 km.

5. ACKNOWLEDGEMENT

The author is grateful to several CEGB colleagues for help with the wind turbine blade throw calculations over a long period of time.

This work was carried out at the Central Electricity Research Laboratories and is published by permission of the Central Electricity Generating Board.

6. REFERENCES

Abramowitz, M., Stegun, I.A., 1965, 'Handbook of Mathematical Functions', National Bureau of Standards

Eggwertz, S., Carlsson, I., Gustafsson, A., Linde, M., Lundemo, C., Mongomerie, B. and Thor, S., 1981, Safety of Wind Energy Conversion Systems with Horizontal Axis (HA WECS). The Aeronautical Research Institute of Sweden. Technical Note FFA HU-2229

Hess, F., 1968, The aerodynamics of boomerangs, Sci. Amer. Vol. 219, Issue, p. 124

Iverson, J.D., 1979, Autorotating flat-plate wings: the effect of the moment of inertia, geometry and Reynolds number. J. Fluid Mech. Vol. 92, Part 2, 327-348

Macqueen, J.F., Ainslie, J.F., Milborrow, D.J., Turner, D.M. and Swift-Hook D.T., 1983, Risks associated with wind-turbine blade failures. IEE Proceedings, Vol. 130, Pt. A, No. 9, December 1983

Milborrow, D.J. (1982) Windpower in the UK electricity supply industry. Electronics and Power, 29, 665

Milne-Thompson, L.M., 1966, Theoretical Aerodynamics, 4th Ed. Macmillan

Proctor, M.V., Turner, D.M., 1984, RISKIT - A Program for calculating the spatial distribution of blade fragments thrown from a wind turbine

Redman, G.M., et al. 1978, Wind field and trajectory models for Tornado-Propelled Objects. EPRI No. 748. Final Report

Sheldahl, R.E., Klimas, P.C., 1981, Aerodynamic characteristics of seven symmetrical aerofoil sections through 180° angle of attack for use in the aerodynamic analysis of vertical axis wind turbines, Sandia Labs., Alburquerque

Synge, J.L., Griffith, B.A., 1959, Principles of Mechanics, 3rd Edition, McGraw-Hill, p 319

Sorensen, J.N., 1984, On the Calculation of Trajectories for Blades Detached from Horizontal Axis Wind Turbines. Wind Engineering Vol. 8, No. 3, 1984

Turner, D.M., 1987, A Monte Carlo Method for Determining the Risk Presented by Wind Turbine Blade Failures

APPENDIX A

RIGID BODY DYNAMICS

The Euler equations for the rotation of a rigid body are central to the solution of the fragment problem. The aerodynamic forces on the right hand side of equation (3.4) are small in comparison to the gyroscopic terms. Thus the solution of the Euler equations with no forces acting is discussed below, these equations are:

$$A \frac{d\omega_1}{dt} - (B-C)\ \omega_2\omega_3 = 0$$

$$B \frac{d\omega_2}{dt} - (C-A)\ \omega_1\omega_3 = 0$$

$$C \frac{d\omega_3}{dt} - (A-B)\ \omega_1\omega_2 = 0$$

Apart from the trivial solution $\omega_1=\omega_2=\omega_3=0$ a basic solution in terms of the Jacobian Elliptic functions (Abramowitz and Stegun, 1965) sn, cn, dn can be obtained. There are two integrals of the above equations leading to constants of motion, these are

$$2T = A\omega_1^{\ 2} + B\omega_2^{\ 2} + C\omega_3^{\ 2}$$

$$h^2 = A^2\omega_i^{\ 2} + B^2\omega_2^{\ 2} + C^2\omega_3^{\ 2}$$

with T the kinetic energy and h the modulus of the angular momentum vector. Suppose the moments of inertia satisfy $A<B<C$ then it can be shown Synge and Griffith (1959) that solutions of the following form exist

$$\omega_2 = \beta\ \mathrm{sn}(pt)$$

If $\quad h^2 < 2BT \qquad\qquad$ ... (A.1)

$$\omega_1 = \alpha\ \mathrm{dn}\ (pt) \qquad \omega_3 \quad \gamma\mathrm{cn}\ (pt)$$

if $\quad h^2 > 2BT$

$$\omega_1 = \alpha\ \mathrm{cn}\ (pt) \qquad \omega_3 = \gamma\mathrm{dn}\ (pt)$$

where α,β,γ,ρ are positive functions of the constants A,B,C,h,T. Sketches of the Elliptic functions are given by Abramowitz and Stegun. sn and cn are oscillatory whereas dn gives rotation in only one direction. The above solutions are mathematical statements of the fact that the body (in the absence of forces) will rotate about either the major axis of inertia (if $h^2 > 2BT$) or the minor axis (if $h^2 < 2BT$). The above two solutions correspond to very different motions. Initially the blade rotates around its major axis of inertia. Although the aerodynamic forces are small at each time step in comparison to the gyroscopic forces they can have a dramatic effect upon the motion if they change the sign of $h^2 - 2BT$ during the trajectory since h,T are only constant in the absence of forces. If h^2-2BT becomes negative along the trajectory then the blade will start to rotate rapidly about its pitching axis.

Table 1: Parameter Values Used in the Tests Peformed Here

Parameter	Value
Moments of inertia about body axes one and two respectively.	5.805×10^4 kg m^2 6.700×10^5 kg m^2
Initial throw velocity along inertial co-ordinates, corresponds to a tip speed of Mach 1.	$170/\sqrt{2}$ ms^{-1} 0 ms^{-1} $170/\sqrt{2}$ ms^{-1}
Wind velocity along each of the inertial coordinate directions.	0.0 ms^{-1} 40.0 ms^{-1} 0.0 ms^{-1}
Initial throw position of the c of g of the blade relative to inertial axes.	$15/\sqrt{2}$ m 0 $60-15/\sqrt{2}$ m
Length of each aerofoil considered.	13 m
Chord of each aerofoil considered.	2.8 m

Initial throw orientation given by

$$R^{(o)} = \begin{bmatrix} 1/\sqrt{2} & 0 & 1/\sqrt{2} \\ 1/\sqrt{2} & 0 & -1/\sqrt{2} \\ 0 & 1 & 0 \end{bmatrix}$$

Table 2: Results Obtained Using the Six Degrees of Freedom Model Described in Section 3

	Initial Angular Velocity Rad/s	Angular Velocity On Impact Rad/s	Axis of Rotation on Impact	Time Step s	Moment of Inertia	Drag Parameter m^{-1}	Time of Rotation Change s	X Coord of Impact m	Y Coord of Impact m	Range m
Case 1	-3	-2.9	1	0.01	0.5A+B	0.001	0.5	649	722	971
Case 2	-3	-3.0	1	0.005	0.5A+B	0.001	0.6	632	685	932
Case 3	-3	-5.0	1	0.002	0.5A+B	0.001	0.6	665	687	956
Case 4	-10	-3.6	3	0.005	0.5A+B	0.001	N/A	445	573	726
Case 5	-10	-9.6	1	0.001	0.5A+B	0.001	0.5	594	390	711
Case 6	-10	13.3	1	0.0002	0.5A+B	0.001	0.5	615	535	815
Case 7	-10	-11.6	1	0.001	0.05A+B	0.001	0.5	561	379	677
Case 8	-10	6.71	1	0.001	0.95A+B	0.001	3.5	663	640	922
Case 9	-10	11.6	1	0.001	1.2A	0.001	0.5	527	548	760
Case 10	-10	-9.9	1	0.001	0.8A	0.001	0.5	599	370	704
Case 11	-10	11.8	1	0.001	1.2B	0.001	0.5	536	512	741
Case 12	-10	13.4	1	0.001	0.8B	0.001	0.5	489	507	704
Case 13	-3	-5.9	1	0.001	0.5A+B	0.002	0.5	696	583	908
Case 14	-10	12.5	1	0.001	0.5A+B	0.002	0.5	475	488	681

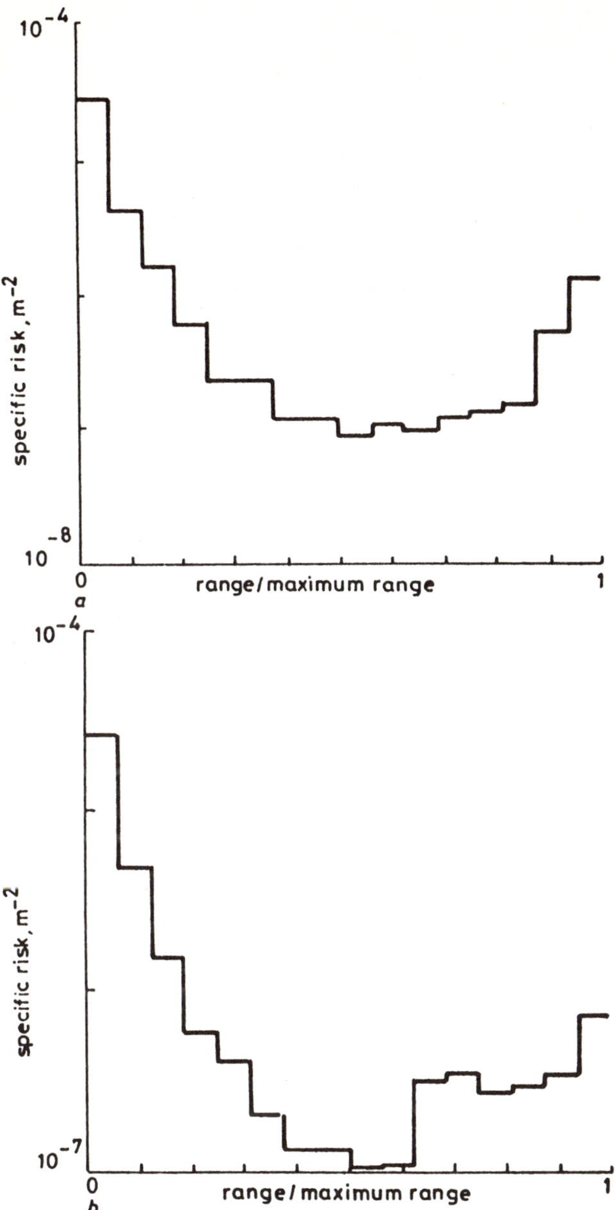

Fig. 1 *Releases in extreme conditions with zero lift*

a Tip fragment
Release velocity = 340 m/s, wind speed = 40 m/s, hub height = 61 m, release radius = 46 m, drag = 4×10^{-3}/m, lift = 0, maximum risk = $2.776 \times 10^{-5}/m^2$, maximum range = 795 m.

b Complete blade
Release velocity = 170 m/s, wind speed = 40 m/s, hub height = 61 m, release radius = 23 m, drag = 2×10^{-3}/m, lift = 0, maximum risk = $2.648 \times 10^{-5}/m^2$, maximum range = 799 m.

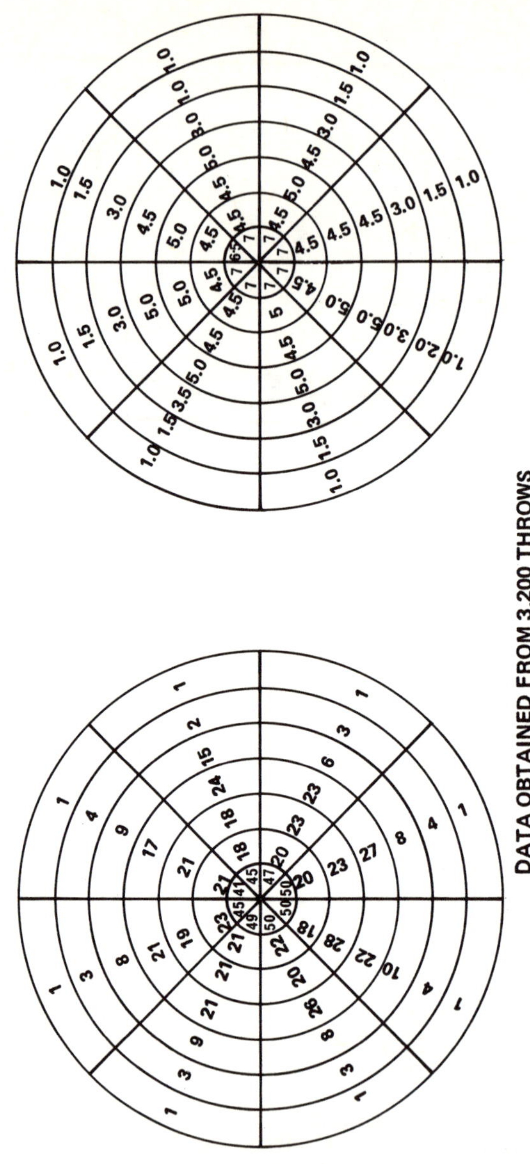

DATA OBTAINED FROM 3,200 THROWS
RADIAL DISTANCE OF EACH SEGMENT = 100 m
SEE SECTION 3 FOR DISCUSSION OF RESULTS

FIG. 2 THROWS AND CONFIDENCE INTERVALS AROUND A WIND TURBINE

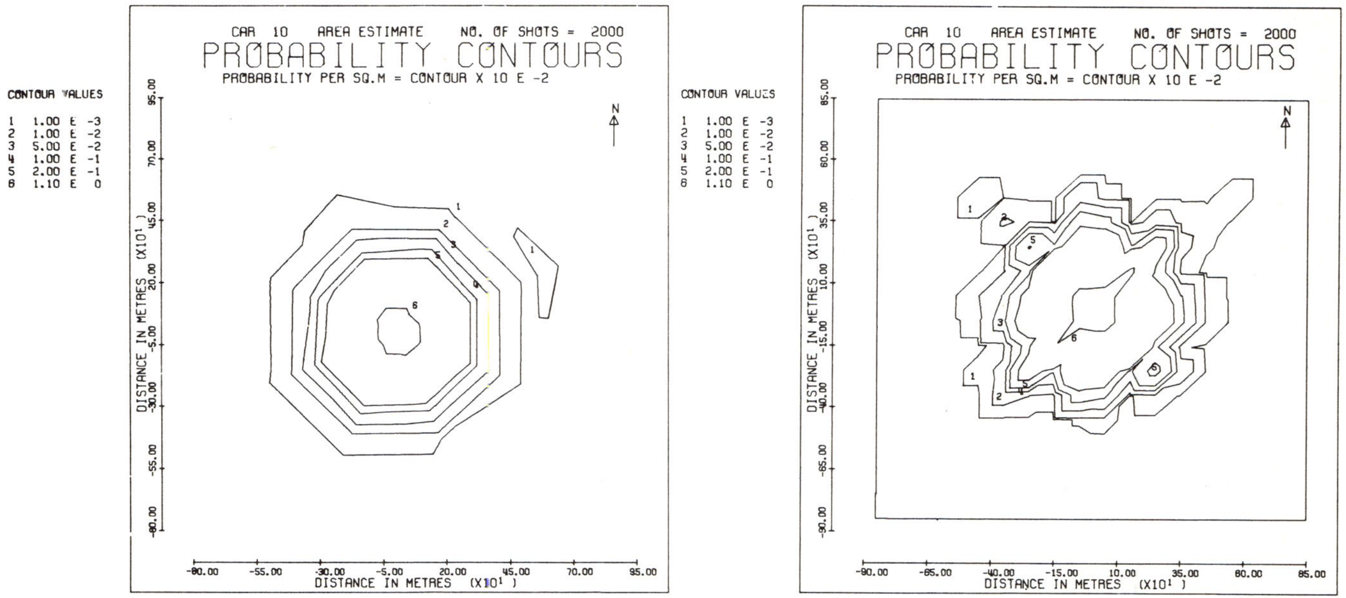

FIG. 3 PROBABILITY CONTOURS (FOR HITTING AN OBJECT OF HEIGHT 1.8 m AND AREA 1.0 m)

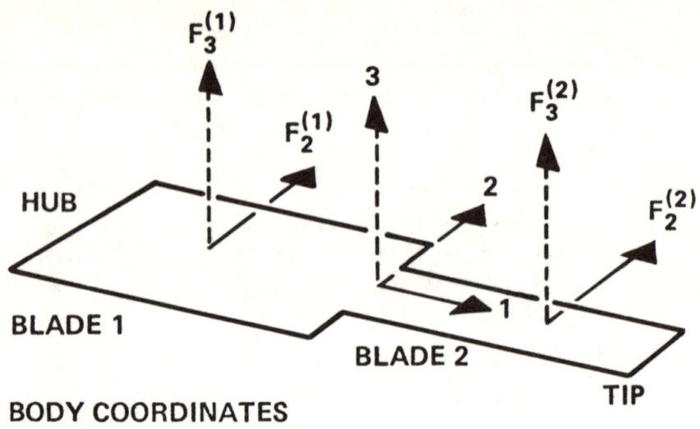

BODY COORDINATES

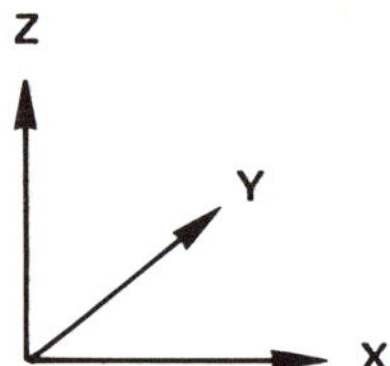

FIXED SPATIAL (INERTIAL) COORDINATES, Y DENOTES THE WIND DIRECTION AND Z VERTICALLY UPWARDS

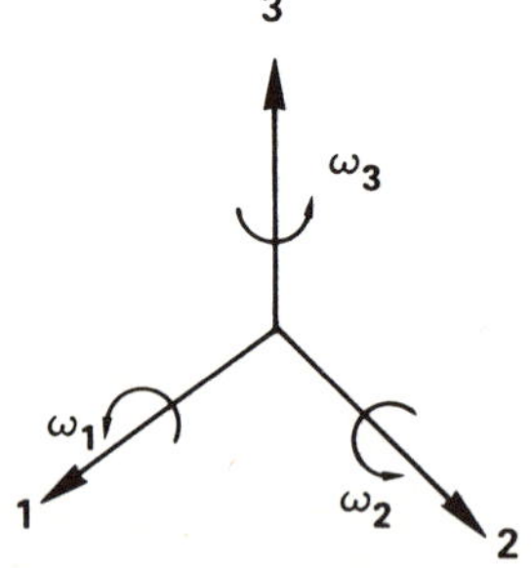

CONVENTIONS FOR POSITIVE ANGULAR VELOCITIES ω_i, i = 1, 3

FIG. 4 COORDINATE SYSTEMS

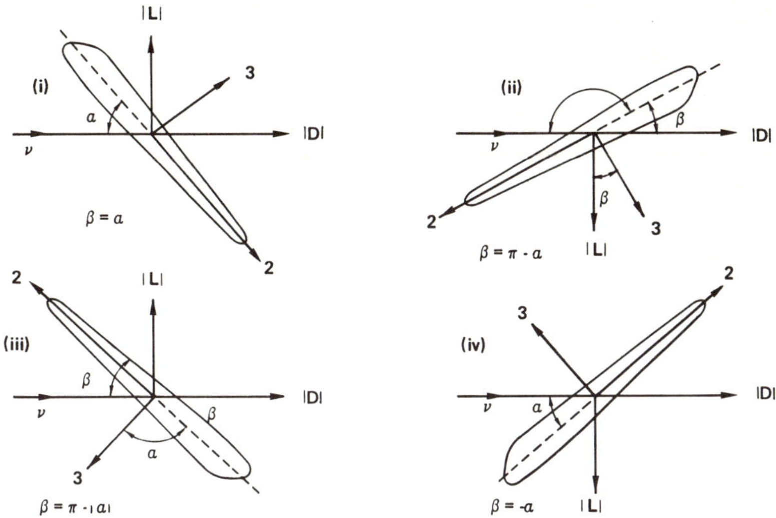

(a) SPAN–WISE AXIS COMING OUT OF THE PLANE OF THE PAPER

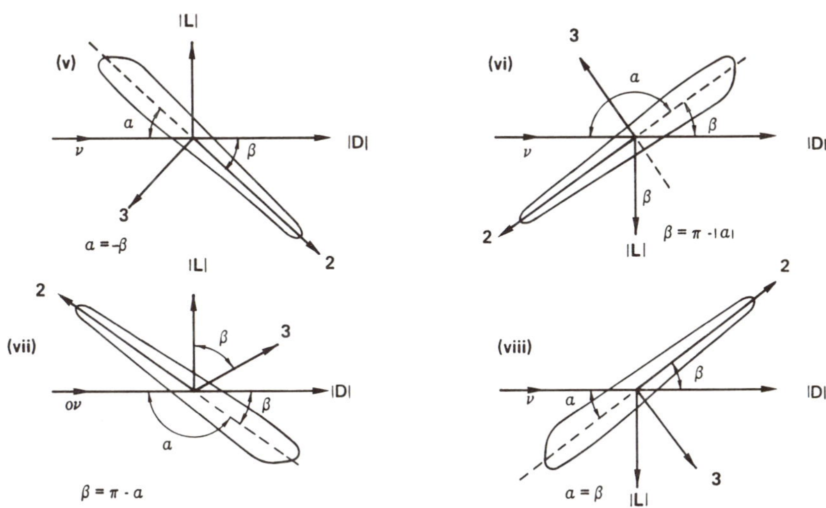

(b) SPAN–WISE AXIS GOING INTO THE PLANE OF THE PAPER

FIG. 5 AERODYNAMIC FORCES ON A 2–DIMENSIONAL AEROFOIL

Chapter 14

Wind energy development in Greece

J. L. Tsipouridis

SYNOPSIS This paper presents PPC's activities in wind energy development in Greece, focusing on problems and difficulties connected with environmental aspects. It also gives a brief account of the fundamentals contained in the relevant Goverment Act.

1. INTRODUCTION

Greece is blessed with an abundance of sunshine while its wind potential , especially in the islands is more than promising.These two facts coupled with the energy crisis and the resulting high cost of imported fuel dictated the need for a national energy policy which would take under consideration local energy sources,integrating them into the national production system, to as high a degree of penetration as technology would allow.

PPC as the sole producer of electricity in Greece and subsequently the medium for the implementation of energy policies, developed an active interest in the exploitation of renewable energy sources. The selection of the testing ground was easy. The Aegean islands had both high wind potential (average wind speeds ranging from 7-11 m/s)and high cost of electricity production from the autonomous diesel stations, as these islands are not interconnected to the national grid . The results and conclusions, where possible, from PPC's experience in this field are presented herebelow.

2. THE INSTITUTIONAL FRAMEWORK

The advent of renewable energy forms and wind energy in particular created the necessity for a Goverment Act to cover the rising demand for private installations of wind energy units and to set the safety, planning and legal criteria. The Act and the Ministirial decisions which followed soon after, covered all necessary aspects of the matter and the response of the public (private companies, organisations, local authorities) has been more than promising.

The Act provides that following the granting of permits for installation from the Ministry for Energy and PPC (in cases of parallel operation to the grid), a project can be subsidised by 35-50% of its total budget (which includes a 15% bonus for energy projects and ecologically sound projects), while furthermore low-interest loans are also available to investors. These, coupled with PPC's obligation to buy surplus electricity from private producers, make the investment quite attractive, hence the response previously mentioned.

The Act covers aspects of planning (distances from housing estates roads, nearby buildings etc.) safety and interference (distance from airports, radar, TV transmitters etc.) and protection of areas of special interest. PPC's projects are affected by most of the above, especially in small islands where most of PPC's projects are intended for installation, and where land availability with the required specifications (including good wind regimes) is rather low.

In addition, and as Greece is covered with archaeological sites, PPC has to tread very carefully to keep as far away as possible from such (and similar) protected areas.

Thus while PPC is making a point of minimising the impact of its wind energy exploitation programme in the Aegean islands, and in addition the use of windgenerators is an ecologically sound solution to the energy problem, it is self-evident that the island environment is imposing considerable restraints on PPC's efforts, rendering the whole operation more difficult.

3. THE ENVIRONMENT - GREEK ISLANDS.

Greeks have always used wind energy for their energy needs in the islands and the windwills which remain, scattered in the Aegean, bear witness to this fact. Thus for us, the new technology was the logical evolution that blended with tradition and environment.

It has already been pointed out that the two factors which primarily rendered the installation of wind generators in the Aegean , islands attractive were the high wind potential and the high cost of electricity production from the autonomous diesel stations. However this choice had its negative aspects which made PPC's work that much more difficult.

The vast majority of these islands which are a haven for tourists, do not possess the infrastructure required for the job. To begin with, access to the islands is difficult if not impossible from October to March each year. Furthermore access roads to windy sites are narrow, winding and generally unsuitable for vehicles. More often than not road works are necessary to make it possible for trucks and cranes to get through, and in all cases cranes have to be ferried over.

The identity and colour of the Greek islands that we know, is partially based on such aspects as the above, and PPC is very sensitive to the fact that its impact on the islands' environment should be the minimum possible. Consequently and in connection with the restrictions imposed for protected areas (archaeological sites, forests, lakes etc.)for safety, interference, and planning (which for the islands icludes keeping well away from villages for aesthetic reasons), site selection is of the outmost importance. Moreover if the technical problems connected with the lack of infrastructure and the criterio for good wind potential are included in the calculation then site selection becomes an "art" in itself. Finally PPC has to ensure that it does not create irreversible changes to the environment, and so it makes a point of having all its civil works blend with local architecture.

Thus taking into consideration all the above parameters - environmental, scientific and technical-PPC has planned and is executing an extensive wind energy programme aiming at utilising the available high wind potential as best as possible.

4. PPC'S WIND ENERGY PROGRAMME

Wind measurements have shown that the average wind speed in the Central Aegean islands is well above 7 m/s with typical sites ranging from 8-10 m/s and extreme cases of up to 12 m/s.

North and South Aegean islands have an average wind speed of about 6 m/s .These figures, indicating the high wind potential available, something that was known to locals from experience, were the original stimulus, which coupled with the high cost of electricity production in the islands weighed heavily in favour of the decision to begin in earnest a wind energy exploitation programme in the Aegean islands.

PPC's first experience came from the Kythnos Wind Park (5X20 KW units), the result of cooperation between Greece and W.Germany. The plant was experimental in nature and from its operation in parallel to the islands diesel station and a 100 KW P/V station, PPC acquired the necessary knowledge that was required to move to other projects.

The Kythnos Wind Park was followed up with two EEC demonstration projects one in Mykonos (108 KW/stall) and one in Karpathos (175KW/pitch). The choice of these two medium sized units (MICON (Denmark) and HMZ (Belgium) respectively) was intended to give us insight in the two different technologies in view of future programmes.

Two more demonstration programmes were accepted by EEC in the following years which differed fundamentally from the previous ones in that the wind generators (100KW horizontal axis and 100KW vertical axis) were of Greek conception, design and manufacture. These projects were intended to initiate manufacturing activity at home while simultaneously testing on the same site windgenerators of different technology.

Finally the two most recent demonstration projects concern the installation of a 400KW James Howden machine in Mykonos (which proved to be an excellant site with average wind speed of 11 m/s) and the installation of 350KW vertical axis machine (SIEMENS) in Andros (with average wind speed of above 9 m/s). Additionally the Kythnos MAN 5X20KW units will be replaced by new improved MAN 5X30KW units, as the old ones have run down.

However the big opportunity for wind energy exploitation was given with the EEC financed HORS QUOTA programme which concerns the installation of 25, 100KW units and 24, 55KW units in the Aegean.

More specifically this programme totaling 3.9 MW of installed capacity is spread in the Aegean islands (shown in figure 1) as follows:

Limnos	6X55KW &	6X100KW
Samos		9X100KW
Chios		8X100KW
Thira		3X100KW
Ikaria	7X55KW	
Karpathos	5X55KW	
Samothraki	4X55KW	
Kea	2X55KW	

The next big step in wind energy exploitation will be taken within the framework of the EEC M.I.P and Valoren programmes, for which PPC is planning the following wind parks (total capacity 13.0 MW).

1. Samos	2.0MW
2. Chios	2.5MW
3. Lesbos	2.0MW
4. Andros - Tinos	1.5MW
5. Euboia	5.0MW

The size and technology of the units to be used will be decided at a later stage.

In conclusion it should be said that PPC has taken onto itself the responsibility of introducing and expanding, to a considerable degree, wind energy in Greece, proving both its technical and economic viability . However, and in spite of the extent of its programmes, PPC is very sensitive to the enviromental issue and is constantly aiming at minimising the impact of its activities on the Aegean islands.

One can only wish that people and organisations promoting other energy sources would be as careful and as sensitive about the environment as are all those associated with wind energy activities in general, in spite of the obvious fact that wind energy by definition blends with the ecological cycle.

BULGARIA
JUGOSLAVIA
HELLAS
TURKEY
N
Samothraki
Limnos
Lesvos
2MW
Skyros
Chios
2.5MW
Euboia
5MW
Andros
1.5MW
Tinos
Kea
Kythnos
Mykonos
Ikaria
Samos
2MW
Arki
Thira
Antikythira
Karpathos
KRITI
Aghia Roumeli
Gavdos
LEGEND
PV PLANT (OPERATING)
PV PLANT (ON GOING)
WIND PARK (OPERATING)
WIND PARK (ON GOING)
WEC (OPERATING)
WEC (ON GOING)

FIGURE 1

Chapter 15

An aerogenerator in a residential environment—the experiences of residents and motorists in the energy park, Milton Keynes

L. Cousins and C. Ledward

SYNOPSIS

This paper sets out the methodology and main findings of a research project investigating the social impact of an aerogenerator situated in a residential area of Milton Keynes.

Views were sought from local residents and passing motorists, and together with observational research and physical noise monitoring, information was provided on the impact of the aerogenerator.

1. BACKGROUND

The Aerogenerator is located within the Milton Keynes Energy Park. This Park is an area of the city that is being planned to combine all policies which promote energy efficiency consistent with creating an attractive environment in which to live and work, and make practical use of new developments in energy and communications technology, drawing on the experience already gained in Milton Keynes.

The 300 acre site is being developed over a seven year period and comprises employment areas, housing, parkland and a range of community facilities including schools, shops and exhibition facilities. It will eventually house 3000 people and provide employment for about 2000.

The Aerogenerator, along with a series of photovoltaic arrays provides electricity on a commercial scale to 9 homes in the Energy Park. This co-generation scheme has been promoted and developed by Solarpak Limited and has attracted financial support from the European Commission.

The Aerogenerator itself is located at the eastern edge of Shenley Lodge grid square, closeby to one of the city's main thoroughfares. To the west of the Aerogenerator is an area of sale housing, with some dwellings as close as 20m.

Milton Keynes Development Corporation undertook research, investigating noise disturbance to local residents and the potential distraction to passing motorists of the Aeorgenerator. Funding for this work was provided by the Energy Technology Support Unit, Harwell, England.

2. OBJECTIVES OF THE RESEARCH

The primary objectives of the research were:

- to provide a rational evaluation of the aeorgenerator's location in Shenley Lodge, Milton Keynes;

- to provide criteria for evaluating future sites for aerogenerators in a residential environment.

- to present a critique of the methodology used and to develop further techniques for assessment of noise and intrusion factors of aerogenerators, where their proposed location is in a domestic environment.

3. RESEARCH METHODS

Three research methods were employed:

- Resident Feedback : A survey consisting of 103 face-to-face interviews with householders living on the Energy Park and closest to the aerogenerator.

- Driver Feedback : A roadside interview survey of drivers passing the aerogenerator site.

- Observational Research : A survey consisting of observations of driver behaviour at the aerogenerator site and its approaches.

The British Market Research Bureau were employed to carry out the interviews with residents and drivers. MKDC was responsible for the overall management of the research and for the observational study.

4. MAIN FINDINGS

The main findings from the research are as follows:

Resident Feedback (All of Sample)

- Overall, residents have a very positive view of life in Shenley Lodge, appreciating its quietness and its appearance as well as the low heating costs of their homes. The main drawbacks are the lack of facilities and the amount of building work still going on at the time of interview.

- Unprompted, only 4% of residents mentioned the aerogenerator as a problem.

- Only 7% of residents wanted the aerogenerator taken away altogether, but a further 26% wanted it moved to a different place away from the houses. 63% were happy for the aerogenerator to stay where it is and 4% didn't know.

- Noise is the main concern with the aerogenerator. When asked directly, 23% of residents said they or their family were disturbed by the noise.

- Concerns about its safety were expressed by those living closest to the aerogenerator, but generally it was thought to be a safe structure.

- The noise produced by the aerogenerator is considered to be continuous when it is operational and more likely to be described as 'whirring' or 'whining' by the 23% who were disturbed by it.

- The aerogenerator is not generally thought to affect the value of property in Shenley Lodge and as many people thought the presence of the aerogenerator would make their home easier to sell, as those who thought a sale would be more difficult.

- Those living nearest the aerogenerator were asked what they had been told about the aerogenerator before they moved in. Only 3 out of 39 residents were told anything by the selling agents.

<u>Resident Feedback</u> (The 15 Households Living Within 130m of the Aerogenerator)

- Residents within approximately 130m of the aerogenerator are those most disturbed by it. Beyond 130m, levels of disturbance fall away quite sharply.

- Disturbance was reported by 10 out of 15 residents and the impact of the aerogenerator is sufficiently serious that thse 10 residents want the aerogenerator moved.

- Worries about safety are held by 9 out of the 15.

- Night-time noise disturbances are worse than day and evening but during the day and evening over half of the residents are disturbed.

<u>Driver Feedback</u>

- A high proportion (67.1%) of the 965 drivers surveyed passed the aerogenerator at least 2 to 3 times a week. Yet 57% remember seeing the structure <u>unprompted</u>.

- After prompting, the aerogenerator is noticed by 90% of drivers but they pay little attention to it.

- 21% of drivers thought the aerogenerator distracting and 19% thought it so distracting as to be the likely cause of an accident.

- 70% of drivers found the aerogenertor a useful landmark.

Observational Research

- The observed effect of the aerogenerator on driver behaviour is marginal.

- The observational research did not identify dangerous driver behaviour linked to any distraction caused by the aerogenerator.

5. CONCLUSIONS

The above findings are based on individual residents' perceptions. The conclusion that the noise disturbance for this type and size of aerogenerator was critical within a 130m distance band needs qualifying by recognising that individual resident thresholds of disturbance are not uniform. This was clearly indicated by the research findings which showed that a minority of residents living within 130m were not disturbed whilst a handful of residents living between 300m and 400m away from the aerogenerator were disturbed.

This issue of differing perception thresholds was partially overcome by supporting the attitudinal research with a physical noise monitoring programme. This monitoring confirmed the findings of the attitudinal research and the noise disturbance was intrusive within the critical distance band identified (130m).

As a consequence of the perceived and identified levels of noise disturbance it was recently decided that the aerogenerator should be relocated in a more suitable location, one which was at least 150m away from the nearest residential building.

SPD/LC/CL/JR/lc3
21.3.88

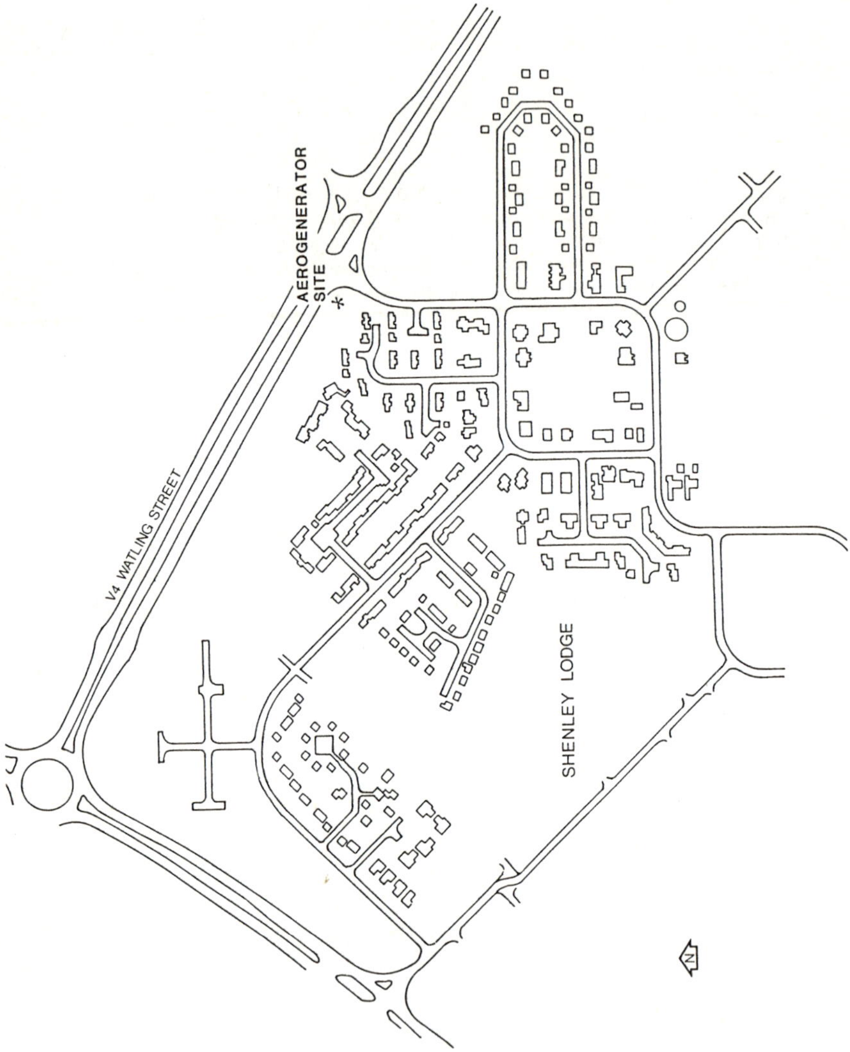

AEROGENERATOR SITE
V4 WATLING STREET
SHENLEY LODGE
N

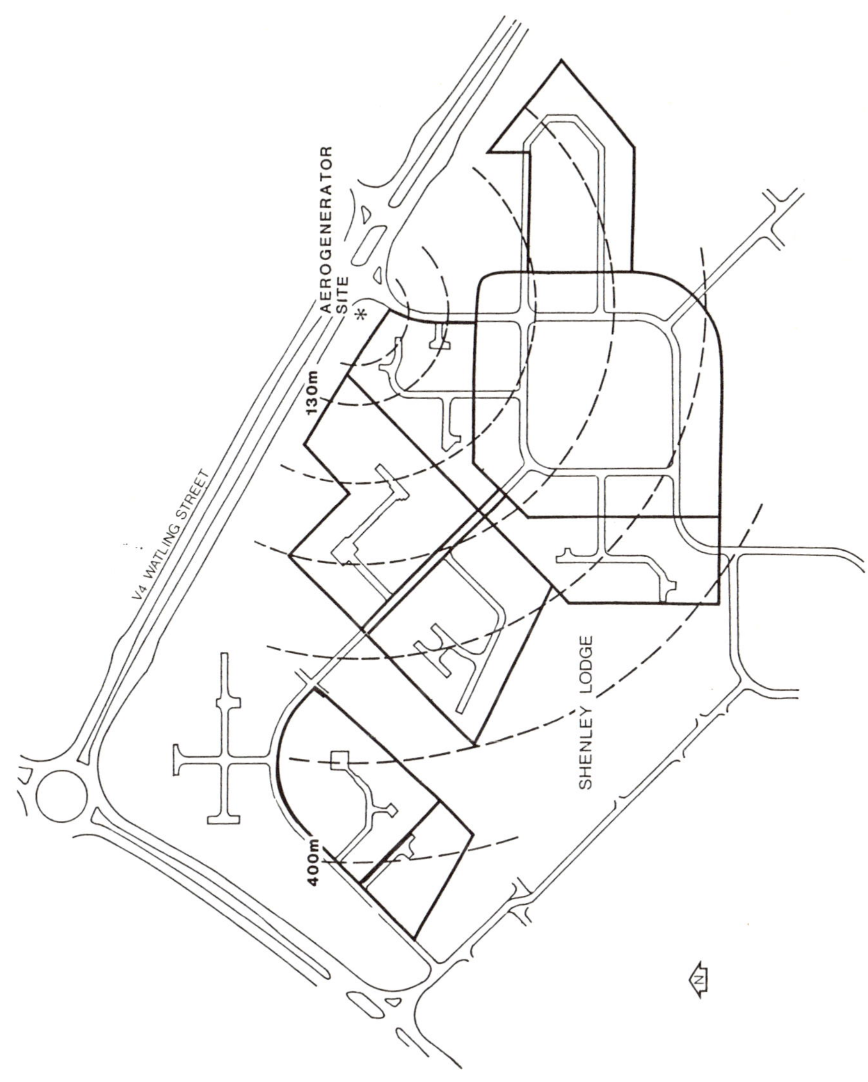
AEROGENERATOR
SITE
130m
400m
V4 WATLING STREET
SHENLEY LODGE
N

Chapter 16

Local attitudes to three existing wind turbine sites in the UK

M. J. Varley, T. D. Davies, C. G. Bentham and J. P. Palatikof

SYNOPSIS

Representatives from about 20 households close to three existing wind turbine sites were interviewed. The three sites are: (a) near Helston, Cornwall, (b) near Ilfracombe, Devon, (c) Burgar Hill, Orkney. The questionnaire was designed to assess attitudes both towards the local wind turbine(s) and to wind energy in general. The sample was not large enough to disaggregate by location.

The survey elicited a generally favourable response to the notion of enhanced exploitation of renewable resources. Although the large majority of those interviewed could hear the local turbine(s) from their property, most regarded the noise level as acceptable, although proximity and prevailing conditions were important. Less than 25% of the respondents claimed that the distance between their household and the turbine site was too small. Television reception interference was not seen as a major problem. The majority of the respondents had neutral or mildly-positive views on the local operation, when asked to summarise their attitudes in terms of 'like' or 'dislike' categories.

The most significant concern was that large numbers of turbines would be needed for a power output comparable with a conventional power station. There were differences in the acceptability of various potential locations. Although the samples were too small for inter-site comparisons, there were clear indications that further studies on site-specific responses are warranted.

1. INTRODUCTION

Generation of electricity from the wind seems to enjoy a generally favourable public image as 'pollution-free' technology. Indeed, it has been promoted by some environmental lobby groups as a way of meeting some energy needs that would involve less damage (or potential damage) to the environment than fossil or nuclear fuel generation. However, the further development of wind power on-shore in the United Kingdom will necessitate the erection of highly conspicuous structures in the landscape. When these are under construction, and during their operation, it is inevitable that there will be a number of environmental impacts. There can be no guarantee that the generally favourable public perception of wind energy will ensure its immunity to public opposition on environmental grounds, once decisions on the siting of major facilities have to be made. Indeed, the fact that many of the technically most suitable sites are likely to be in upland and coastal areas of considerable amenity value virtually guarantees that public acceptability of environmental impacts will be an issue. It is therefore important to assess public attitudes to existing wind power installations.

We start to explore the current views of the public on wind turbines by examining the attitudes of people who have direct experience at three locations: (i) Ilfracombe (North Devon); 250 kW turbine; (ii) Helston (Cornwall); 125 kW turbine; (iii) Burgar Hill (Orkney); 3MW, 250kw and 300kW.

2. MEASUREMENT OF ATTITUDES

(a) The Survey Questionnaire

Question selection was based upon work undertaken by previous researchers in the environmental assessment of wind power (2, 3) and, in particular, public opinion papers (1, 5). Scott (4) presents a detailed discussion of attitude measurement. The most common technique is the self report quesionnaire, particularly of the scale type which is adopted here, although respondents were interviewed in this study.

1. Would you like to see (i) an increase, (ii) a decrease, (iii) no change in the proportion of energy supplied by each of the following:

   A. Coal
   B. Oil/Gas
   C. Nuclear
   D. Wind/Tidal/Geothermal
   (iv) no opinion

2. Apart from your local turbine, do you know of any other wind turbines in Britain?: (i) Yes, (ii) No

3. Can you see the turbine from anywhere inside your house?: (i) Yes, (ii) No, or from your garden?: (i) Yes, (ii) No.
   If (i) to either question, do you notice if it is working or not?: (i) Yes, (ii) No.

4. Can you hear the turbine when it is working?: (i) Yes, (ii) No.
   If (i):

   A. Indoors
   B. Outdoors
   C. Both
   1. All the time
   2. Part of the time

   Is the noise level (i) acceptable, (ii) unacceptable?

5. What is your attitude to the turbine?

   A. Extreme dislike
   B. Dislike
   C. Not bothered
   D. Like
   E. Like a lot

6. Do you think that TV reception is affected by the turbine?: (i) Yes, (ii) No.

7. Using the distance from your house to the turbine as a measure of minimum distance for people to live from the turbine, would you say the distance is:

   A. More than adequate (you would live closer)
   B. About right
   C. Too large

8. Assign values 1-3 on a scale of importance to the following disadvantages:

   A. Unreliable, since wind does not always blow
   B. Expensive to research and build
   C. Spoils the scenery
   D. Noisy
   E. Interferes with TV reception
   F. Disturbs wildlife
   G. Large numbers are needed to provide the same output as a Coal or Nuclear power station

9. Assign values 1-3 on a scale of importance to the following advantages:

   A. Clean; does not emit pollution
   B. Safer
   C. Provides cheap electricity
   D. It will never run out
   E. It will make supplies of coal and oil last longer
   F. Provides local employment

10. As wind power develops, more sites for turbines will be needed. Would you find the following sites (i) acceptable, (ii) unacceptable?

    A. Near the coast
    B. In a line along a road (Note: can be in a line along a road near the coast etc.
    C. In a line along a ridge of hills
    D. In an area of scenic beauty
    E. In a largely man-made landscape, such as an agricultural region (Note: may also be area of scenic beauty)
    F. In a cluster
    G. Offshore

11. Would you prefer to see (i) a group of wind turbines, or (ii) one on its own; (iii) not bothered.

12. In a group of wind turbines, would you prefer to see (i) a few larger ones, (ii) more smaller ones; (iii) not bothered.

13. Would you be interested in owning part of a turbine in order to reduce electricity bills?: (i) Yes, (ii) No, (iii) Maybe.

14. Would you like more information about the turbine and its operation: (i) Yes, (ii) No

15. Sex (i) M, (ii) F
    Age (i) <44, (ii) 45-59, (iv) 60+
    How long resided in locality

(b) The Samples

The intention was to interview representatives from households in the immediate vicinity (or within potential visual or noise contact) of the turbine. The very different character of each site (see section 3) precluded the adoption of a sampling regime for all the sites; consequently we can consider the sampling methodology site-by-site.

(1) Ilfracombe: The respondents were chosen on the basis of alternate houses, from four small settlements closest to the turbine: Lincombe (~700m), Higher Slade (~275m), Lower Slade (~900m) and the outer fringes of Ilfracombe (Langleigh Park and Torrs Park area) (~1km) (See Figure 1). Respondents from 22 (out of a total of 45) houses in these areas were interviewed.

(2) Helston: In this case, the sample included all households within 1km of the turbine. There was considerable spatial variability in the degree of potential impact because of the nature of the local topography (Figure 2). For example, visual intrusion was greatest in the region to the southwest of the turbine, around Lower Treleggan and Trebarvah Woon.

(3) Orkney: Representatives from all houses within potential impact distance, defined in this case as within a distance of 2km, were interviewed.

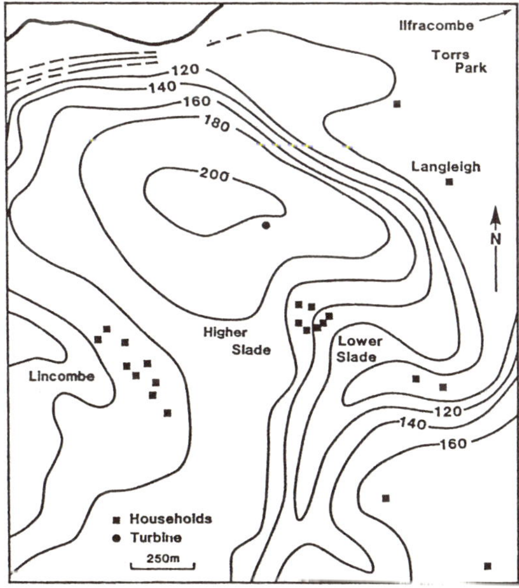

Figure 1 Location of the Ilfracombe turbine. Contours are in metres and the contours <100m are not shown to the north and north-east. The coastline shows in the extreme north. The locations of households which were included in the survey are shown.

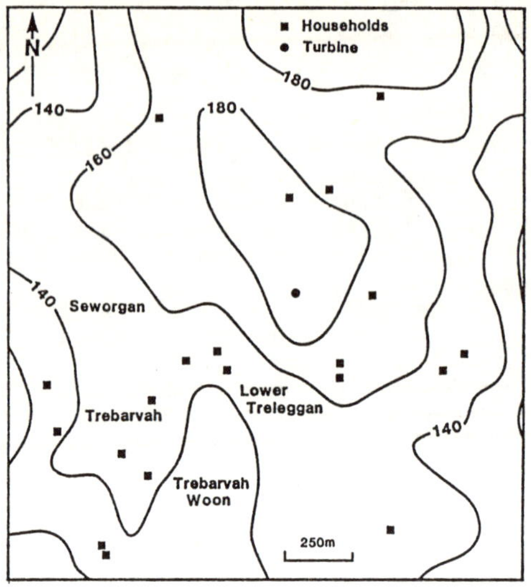

Figure 2 Location of the turbine near Helston in Cornwall

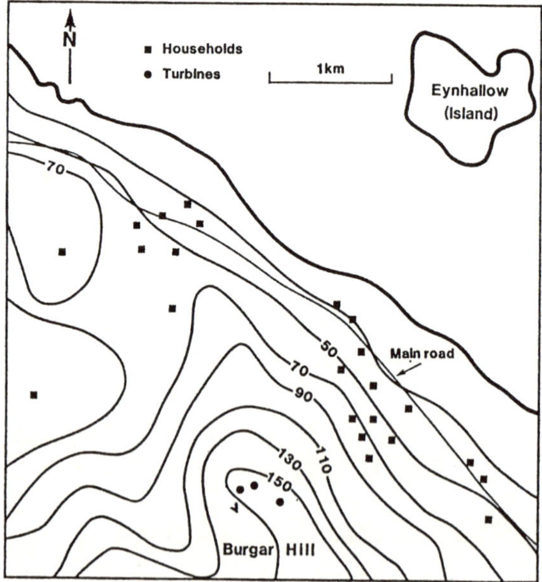

Figure 3 Location of the turbines on Burgar Hill, Orkney

The nature of the sample populations is summarised in Table 1.

	Sex		Age					
	M	F	<44	45-59	≥60	Indigenous	Migrant	Total
Ilfracombe	11	11	10	6	6	3	19	22
Helston	12	9	6	8	7	11	10	21
Orkney	10	13	8	8	7	17	6	23

The relatively great average age of rural populations is reflected in the sample population in all three regions. The most marked difference in the samples from the three areas relates to the indigenous/migrant ratio (migrants were defined as people residing in the area for <10 years): the vast majority of respondents of Ilfracombe were relative newcomers, whereas the greater number of respondents in the Orkney survey were native Orcadians. This respondent structure is of some interest, since Newby et al. (1978) showed that in-migrants to rural areas are much more concerned about environmental and amenity issues than is the 'local' population. A particular conflict they pinpointed is that some rural development schemes are welcomed by the indigenous population because of the potential employment opportunities, but are opposed by in-migrants who wish to retain their surroudings in the same form which frequently played a large part in their decision to move to the area.

3. THE SITES AND TURBINES

(i) Ilfracombe: The 3-bladed, 25m diameter, 250kw turbine was erected in 1984. It is mounted on a slender, 25m tall, tapered solid steel tower. It was operating during the survey period, although it has experienced some operating difficulties and was closed down for five months before the survey. The turbine is located about 250m from a 30m tall television mast, near the top of a steep coastal hill but near the top of a southeast-facing rentrant valley (at about 210m altitude), being hidden from view of the centre of the town of Ilfracombe (1.5km to the northeast) behind a steep elongated spur of the hill.

(ii) Helston: This stall-regulated turbine has an output of 125-145kW; the blades are 17.5m in diameter and are mounted on a 15m solid metal tower. The site is midway between Helston and Falmouth on a southwest-facing slope at a height of ~190m. It is close (~1km) to the village of Seworgan and, although invisible from the village, is visible over a large area. Erection was completed in 1986, and the turbine has worked almost continuously since then. It was operating at the time of the survey.

(iii) Orkney: The site of Burgar Hill is a large wind power development in the United Kingdom. There are three turbines on the site. The 250 kw turbine was installed in 1983; the blades have a 20m diameter and the solid cylindrical tower is 16.3m tall, the machine was working at the time of the survey. The 300 kw turbine has 22m blades mounted upon a 22m solid steel cylindrical tower; it was not working at the time of the survey. The 3MW machine stands 75m tall with a blade span of 60m. It has recently been commissioned and was not working at the time of the survey. Burgar Hill stands 160m high, 2km from the coast (Figure 3); the turbines are readily visible over a large area, including adjacent islands, because of the rolling, gently rounded landscape.

4. RESULTS AND DISCUSSION

No attempt has been made to disaggregate the returns by location because of the relatively small size of sample.

Question 1

Energy source.

	Increase	Decrease	No change	No opinion
Coal	11	17	25	23
Oil/Gas	7	3	39	17
Nuclear	6	39	7	14
Wind/Tidal/Geothermal	55	1	1	9

There was a very strong call for an increase in the proportion of electricity supplied from renewable sources. On the doorstep, it was apparent that wind energy was particularly favoured, but that may have been a function of the main focus of the study. However, geothermal sources were mentioned in Cornwall, and tidal generation in Orkney.

Question 2

Do you know of any other wind turbine in Britain, besides your local machine?

	Yes	No
Indigenous	11	20
Migrants	23	12
Total	34	32

This question was asked in order to provide a measure of interest of the respondents' in wind energy. The total response was evenly split; and there is no evidence (Chi-squared test) that the migrants' response is different to the response of indigenous people.

Question 3

	Yes	No
Can you see the turbine from inside your house?	43	23
Can you see the turbine from your garden?	48	18
Do you notice if it is working or not?	57	8

The responses to this question confirmed that most interviewees were aware of the visible presence of the local machine(s) and, interestingly, the vast majority were aware of whether or not it was operating, even if it was not directly visible from their homes.

Question 4

	Yes	No	No opinion
Can you hear the turbine when it is in operation?	48	18	
If you hear it indoors is it?: (a) part of the time	7 (4 with windows open)		
(b) all the time	5		
Is the noise level acceptable?	41	7	18

The return to this question confirms that the majority of the respondents could hear the machine. About 20% could hear the local machine indoors, at least some of the time. Only seven respondents believed the noise level to be unacceptable. Most of the people who found the noise level to be unacceptable lived near the Helston and Ilfracombe turbines and they exhibited the most antagonistic reaction. Proximity to the turbines was a clear factor here, and there was evidence that wind direction and time of exposure were important. A more detailed study is currently being conducted.

Question 5

Do you like or dislike the turbine?

	Extreme dislike	Dislike	Not bothered	Like	Like a lot
Indigenous	2	1	18	9	1
Migrants	4	5	12	12	2
Total	6	6	30	21	3

For the total respondents, the overall opinion was neutral - mild acceptance. There was no statistically-significant difference in the response of the two groups.

Question 6

Do you think that TV reception is affected by the turbine?	Yes	No
	13	53

Question 7

The distance to the turbine is:	More-than -adequate	About right	Too small
	4	47	15

Question 8

	Small disadvantage	Disadvantage	Great disadvantage	Don't know
Unreliable	37	14	15	0
Expensive	30	17	13	6
Spoils scenery	44	12	10	0
Noisy	42	10	14	0
Interferes with TV	52	7	5	2
Disturbs wildlife	61	3	2	0
Large numbers	10	20	25	15

The major disadvantage perceived is the fact that relatively large numbers of wind turbines are needed to provide an output equivalent to a conventional power station. None of the other characteristics suggested were perceived as overwhelming disadvantages by the respondents, although noise, development expenses and unreliability of output were the most common concerns. Spoilation of the scenery, TV interference or disturbance to wildlife were not viewed as disadvantages by the majority.

Question 9

	Small advantage	Advantage	Great advantage	Don't know
Clean	4	4	57	1
Safer	3	18	44	1
Cheap	11	13	40	2
Inexhaustible	1	8	55	2
Conserve fossil	4	19	41	2
Employment	21	25	17	3

Of the characteristics which can be viewed as advantageous, the pollution-free generation of power and the renewable resource were seen as the clearest advantages. Possible provision of local employment was not viewed as advantageous as the other characteristics.

Question 10

	Acceptable	Unacceptable	Don't know
Near coast	50	13	3
Along road	26	38	2
Hill ridge	44	18	4
Scenic area	31	33	2
Manmade-scape	53	10	3
Cluster	52	10	4
Offshore	57	6	3

A clear majority regarded offshore installation as acceptable, although a coastal location was also acceptable for a large majority. Acceptability of locations along a road or in a 'scenic' area was lower. All the other characteristics were acceptable to more of the respondents than to whom they were unacceptable.

Question 11

Would you prefer to see: a group of turbines 22

one on its own 21

('none' (5) or 'not bothered' (18))

Question 12

In a group of turbines would you prefer: few larger ones 24

more smaller ones 21

('none' (5) or 'not bothered' (16))

Question 13

Would you be interested in part ownership:	Yes	No	Maybe
	36	16	14

A substantial majority claimed interest in a financial stake in a local turbine.

Question 14

More information?	Yes	No
	35	31

5. CONCLUSIONS

The survey questionnaire elicited a generally favourable response to the notion of enhanced exploitation of renewable resources. Although overall awareness of other wind turbine sites in the UK was not high, the level of interest in the operation of the local turbine was substantial. Although the large majority of the respondents could hear the local turbine(s) from their property when in operation, most regarded the noise level as acceptable. Most of the respondents had neutral or mildly-positive views on the local operation, when asked to summarise their attitudes in terms of 'like' or 'dislike' categories. Most did not consider television reception to be a problem, and only 15 (out of 66) claimed that the distance between their household and the turbine site was too small.

The most significant concern was that large numbers of turbines would be needed for a power output comparable with a coal-fired or nuclear power station; and the development costs and firmness of the wind resource were also worries. The most significant advantages were reported as the pollution-free technology and the renewable nature of the source.

There were differences in the acceptability of various potential locations. The most acceptable sites were offshore or near the coast, and amongst the least acceptable was a road-side location. Given the choice of a group of turbines or a single turbine at a site, no clear preference was evident; neither was particular size of individual turbines within a group favoured. A clear majority was interested in part-ownership of a turbine, but there was no overwhelming call for further information on wind power developments.

The collected data did not discern any statistically significant differences in the responses of migrants and indigenous people. Although tests for differences between the two groups are specifically mentioned only in the discussion of the responses to questions 2 and 5, data are available for all other responses and they confirm the absence of any significant difference in the reaction of the two groups in the study. What was clear was that the wider opinions expressed on the doorstep, beyond the constraints of the questionnaire, were often couched in forceful terms: the migrants who disliked turbines appeared to harbour strong feelings, whilst many non-migrants showed little concern about, or were sympathetic to the local turbine.

It should be borne in mind, however, that this survey attempted to elicit "average" attitudes from three locations, with different geography, residents and differing times of exposure to wind turbines. A more detailed study of each of the three locations should be undertaken, and further work at two of the sites is currently being conducted. There are indications, from our study, that there are important site differences which should be accounted for. Those migrants who disliked the local turbine, exhibited very strong feelings. This was a characteristic of the Devon and Cornwall sites. Many non-migrants, which were in the majority in Orkney, were much more sanquine and sympathetic to the local installation. Wider surveys, addressing these differential responses in each area, could well confirm the findings of Newby et al. (1978).

The survey has utilised the opinions and perceptions of people living in close proximity to three separate wind turbine installations in the UK. Consequently, although the samples were too small for disaggregation by site, the returns do provide useful information for anticipation of the reactions of residents in future prospective sites, and some insight into the issues which they might regard as important. The conclusions of this study are probably reasonably encouraging for the development of wind energy. In spite of the interviewed individuals representing the first groups of people to be exposed to a new and unfamiliar technological development, the attitudes of these affected people were not strongly antagonistic but could be regarded as being, in the main, mildly supportive.

6. REFERENCES

(1) Carlman I. (1986). Public opinion on the use of wind power in Sweden. Proc. EWEA Conference, Rome, 1986. 569-573.

(2) Miller J.S. (1987). Environmental Assessment of Wind Turbine Generator Project, Susetter Hill, Shetland. Proceedings of BWEA Conference, April 1987, Newby, H., Bell C., Rose D., and Saunders P. (1978). Property, Paternalism and Power, Hutchinson, London.

(3) Rogers S.E. (1976). Environmental effects of wind energy conversion systems. Battelle, Columbus Laboratories, Ohio 43201.

(4) Scott W.A. (1968). Attitude measurement. In The Handbook of Social Psychology. (Eds. Lindzey G. and Aronson E.). Addison-Wesley.

(5) Wolinsk M. (1986) Public acceptance of large WECs in the Netherlands. Proceedings of EWEA Conference, Rome 1986, 587-592.

Chapter 17

Off-shore wind power

D. T. Swift-Hook

1. INTRODUCTION

Off-shore siting is seen as the ultimate solution to environmental difficulties for wind energy. It has always been envisaged that off-shore wind power would be required if and when no more land-based sites were available for wind turbines eg because of environmental objections. It has therefore been natural to think of off-shore wind power following the development of land-based wind farms perhaps ten years later. Events move rapidly, however (Swift-Hook, 1988). It is now accepted that wind turbines can generate electricity more cheaply than any other type of power plant. On 23rd March 1988, Lord Marshall of Goring, Chairman of the Central Electricity Generating Board, told the British Wind Energy Association of the CEGB's intention to build three wind parks, each covering an area of three or four square kilometres, one in Cornwall, one in Wales and one in North Yorkshire.

He went on to explain that, although it may now be possible to generate electricity at a competitive cost on favourable sites where high wind speeds can be expected, the reaction of the public is uncertain and so the CEGB have already decided to build a prototype off-shore wind turbine installation.

These sudden announcements took many people by surprise for two reasons. Firstly, there was wide-spread amazement (and often out-right disbelief) that wind turbines have become competitive. There was then surprise that environmental constraints were thought to be so severe that off-shore siting needs to be actively considered even before the first few wind parks have been built on land.

This paper starts out by describing the recent developments that have made wind energy economic and outlining the environmental problems that will arise on land. These developments and difficulties have not been entirely unforseen and far-sighted teams of experts have been working in many different countries for the last ten years on the detailed engineering problems of off-shore wind farms. This paper describes these studies. They have always tended to be based upon published and therefore out-dated designs of land-based wind turbines and an attempt is made to predict the present costs of off-shore wind parks based upon today's land-based costs.

It is pointed out that any serious development programme for off-shore wind power should include the development of very large machines, such as that designed by Mensforth (1979).

2. COST CALCULATIONS

Cost calculations which show that wind energy is economic on land have been set out in some detail by the author for the British Wind Energy Association (Swift-Hook, 1987) and in the first paper in this volume. A simplified formula for the cost of energy will be sufficient here:

$$COE = O\&M + C\,R\,/\,E$$

where O&M = operation and maintenace costs, p/kWh
C = capital cost, £
R = real annual charge rate (interest plus amortisation), %/a
E = annual energy generated, kWh/a

For the WEG wind farm in California described in the first paper of this volume, the true cost of energy based upon the first year of operation is 5.5 c/kWh (taking C = $8,000,000, E = 11,500,000 kWh/a, R = 7.1 %/a and O&M = 6 c/kWh), or around 3 p/kWh. Large wind farms could be built in Britain for two-thirds of the cost (per kilowatt) and operation and maintenance costs should be halved. In Britain, the WEG wind farm would generate at less that 2 p/kWh.

It is a complicated matter to calculate the value of electricity to a power system since that depends upon many factors including the particular plant mix of the individual power system itself. However, Electricity Boards in the UK are required to publish the rates at which they are prepared to buy privately generated electricity to cover their avoided costs and those figures therefore should include all of the complexities. They are typically around 2.7 p/kWh for wind. The margin between the generation costs of around 1.7 p/kWh and the value of the electricity to the power system of around 2.7 p/kWh is considerable. For other utilities in other countries, the figures will differ in detail but electricity costs tend to be broadly similar throughout Europe and the margin is sufficiently great to cover most countries. Wind energy's claim to be the cheapest method of generation available is well substantiated by the performance of the WEG wind farm.

3. ENVIRONMENTAL PROBLEMS AND LEGAL CONSTRAINTS

Electrical power utilities naturally want to install the type of power plant that will generate the cheapest electricity. Wind farms now provide the most economical method of generation and so active consideration is being given to their wide-spread use. Questions are still being asked about technical and financial matters such as reliability and the costs of operation and maintenance and answers to many of those questions are emerging from the wind farms already in service. But by far the strongest reactions from the utilities, and the biggest question-marks they are now raising, are to do with environmental problems and siting considerations (Swift-Hook, 1988).

These environmental problems are the subjects of this volume. Wind farms cover considerable areas of countryside and although they occupy only a tiny fraction of the land (only a per cent or two), their presence will be noticeable over the whole area. If they make a noise, they may be heard and if they interfere with radio or television reception, there will be objections. They will certainly be visible at quite a distance and the public will need reassurance about many other matters : no one wants to be hit by a flying wind turbine blade. Off-shore siting of wind farms would avoid most of these difficulties.

Costs will clearly be greater off-shore both for construction and for operation and maintenance, but the wind blows much harder off-shore and so detailed studies are needed to determine the over-all economics.

The resource is considerable for those countries with substantial coastlines. In the UK, for example, national studies considered water depths greater than 10 m and identifed "probable" off-shore areas sufficient to generate 100% of present electricity production, with a further 50% from "possible areas. Complementary CEC studies considered shallower waters around the coasts of Europe and found yet further wind energy resources around the UK.

4. WORLD-WIDE OFF-SHORE WIND POWER STUDIES

Many individual countries and international organisations have made substantial and continuing studies of off-shore wind power over the last ten years. These include:

Denmark
Germany
Great Britain
Netherlands
Sweden
United States

Commission of the European Communities
International Energy Agency

The study by the International Energy Agency (IEA) is the broadest of these and brought together many of the other activities. It will therefore serve as an indication of the approaches that have been adopted. It was formally initiated in 1983, after several years of negotiations, under the leadership of the present author and the CEGB. Denmark , Netherlands, Sweden and Great Britain were later joined by the United States under the aegis of the IEA Agreement on Wind Energy Conversion Systems Research & Development.

The overall objectives were to assess the viability of off-shore wind power, to define criteria for an off-shore wind turbine prototype and to produce an outline plan for the design, construction and operation of a complete off-shore wind farm. It was envisaged that the studies might lead to a joint project for the construction of an off-shore prototype and the CEGB had offerred to host such a project as early as 1980. The international project was divided into five major areas , each led by a different participating country, with overall co-ordination provided by the CEGB.

IEA OFF-SHORE WIND POWER PROJECT	Coord : GB
Design specifications for wind turbines	Coord : SW
Conceptual designs for a wind farm	Coord : NL
Structural dynamics under wind and wave action	Coord : US
Wind and wave data	Coord : GB
Generic studies	Coord : DK

The design specifications for individual off-shore wind turbines are co-ordinated by Sweden (Hardell, 1984; 1986a; 1986b), who had already constructed two large, land-based multi-megawatt wind turbines at Maglarp (near Malmoe) and at Naesudden (on the island of Gotland) by the early 1980s. Each of the other countries now has one or more megawatt designs. The distinctive feature of an off-shore wind turbine is the base upon which it mounted and the participating countries have considered a wide range of options including individual sand-fill islands, concrete gravity bases, piled structures (including large monopiles) and floating platforms.

Off-shore wind farm conceptual designs are co-ordinated by the Netherlands (van Iperen, 1980; Koten,1986). Important aspects include the electrical inter- connections and shore connection, access for maintenance (by boat or helicopter) and dock yard/construction facilities which can represent a significant fraction (up to 30%) of the total cost of an off-shore wind farm.

The structural dynamics of an off-shore wind turbine are complicated by low frequency wave forces on the lower part of the tower. The United States are carrying out detailed studies in this area, extending the work on aero-elasticity that is done for land-based machines.

Accurate off-shore wind data are difficult to obtain, particularly when vertical profiles are needed (Bennett et al, 1983). Relatively few meteorological masts or their equivalent exist off-shore and those oil rigs, gas rigs, light vessels and weather ships that are in place are not always fully instrumented for wind energy requirements. Use must be made of fortuitous reports from ships of passage and mariners' visual observations of sea states (and, more recently, of satellite photographs) to supplement the available data. The compilation of off-shore wind data is being co-ordinated by the CEGB.

Finally, the compilation of other generic data on materials and other matters is co-ordinated by Denmark (Maribo Pedersen, 1983).

A substantial number of other major reports and papers have been contributed to all of these studies (eg Burton,1985; Otway,1986; Palutikov et al, 1985)

There is a considerable divergence of views between the participants regarding the technology and the economics (Otway, 1986). British figures, which are based upon old designs of wind turbines that have not been constructed, are double those produced by Sweden, which are based upon existing designs. The Swedish costs look interesting (around 3 p/kWh) while the British ones do not (around 7 p/kWh).

Unfortunately, none of the technology nor the costs in any of the studies reflect the latest developments in land based machines which have produced such dramatic cost reductions in the last year or so. The participants generally now feel a good deal more optimistic than they did earlier in the studies and they suggest that the future generating costs of off-shore wind farms will be around 3 p/kWh and perhaps even lower. They have spent substantial sums on these paper studies (more than £1,500,000) and further amounts do not seem to be available for detailed updating in the light of recent technology.

5. OFF-SHORE WIND POWER COST PROJECTIONS

Many detailed costings have been carried out in the various studies but they all suffer from the difficulty identified by the IEA of keeping up with the latest technology and costs.

As an alternative general approach, it should be possible to use the experience developed in the course of the studies to derive cost ratios for off-shore installations. It is felt that the additional cost of off-shore construction should not add more than 40% or 50% to the capital costs, if most of the construction is completed on land and the wind turbine is transported out to be erected on a pre-formed base. This indicates capital costs of around £900/kW for the construction of off-shore wind turbines in quantity.

Very considerable reductions have been achieved in operation and maintenance costs on land by careful design for remote and automatic operation and for easy and rapid maintenance by unit replacement. This approach is clearly correct for off-shore installations and further savings may be possible. There is a good deal of experience in the off-shore oil industry which shows that operation and maintenance costs are three times those on land. In California O&M costs are around 0.3p/kWh; in Britain on land they should be half that and therefore off-shore O&M costs should be around 0.5 p/kWh .

Winds blow considerably harder off-shore than they do on land (Bennett, 1983), except where there is some topographical acceleration over hilltops. 8.7 m/s or 8.8 m/s is the annual average wind speed in some of the off-shore area investigated around the Britsh Isles. This will give 30% higher annual average wind energy density than the 7.8 m/s sites considered on land in Sections 3 and 5, and the annual average load factor will be around 38% instead of 28%.

Taken together, these various factors give a toal generation cost of 2.4 p/kWh which is less than the purchase price offerred by English and Welsh Electricity Boards. It can therefore be concluded that:

IF off-shore construction costs 50% more than on land and
IF off-shore O&M costs three times more than on land and
IF the wind blows 30% harder off-shore (8.8 instead of 7.8 m/s),
THEN off-shore wind power is already ECONOMIC.

6. OFF-SHORE WIND TURBINE INSTALLATIONS

The various national and international studies have produced a very considerable amount of information and, although a joint construction project no longer seems likely, several countries are planning off-shore demonstrations of individual wind turbines and wind farms.

Several countries have installed wind turbines that are literally speaking off-shore, although they are close to land or connected by a causeway and not regarded as such. The wind turbines along the mole at Zeebrugge are perhaps the most widely known, having appeared on the television screens of the world following the ferry disaster there. Denmark, too, have a harbour installation in Jutland and Germany have a 1 MW unit just off the island of Helgoland. Many countries have wind turbines on coastal or island sites. The CEGB site at Carmarthen Bay comes into this category, and possibly the NSHEB site on Orkney Mainland. The Dutch test station run by ECN at Petten is on the coastal dunes and the Isselmij wind farm is only a few metres behind the sea defences. In Denmark, the two 650 kW Nibe machines are only a few metres from the shore. In all of these cases it can be claimed that the wind regime is off-shore, at least when it blows in particular directions, and that there are some features relevant to off-shore situations. However, crucial features such as wave action and inaccessibility are missing.

Several truly off-shore demonstrations are now planned. The CEGB will build a 750 kW unit off Wells near the Northernmost point of Norfolk. The Netherlands have firm plans for an off-shore wind farm. Denmark is seriously considering two off-shore sites for widfarm projects, one in Aarhus Bay, Jutland and the other off Amager Island near Copenhagen. A consortium has been formed in Sweden to plan and construct an off-shore wind farm off Bleckinge in Southern Sweden.

7. VERY LARGE MACHINES

Most studies to date show that there are large fixed costs associated with an off-shore wind turbine. The cost of providing a base on the sea-bed is not very sensitive to the size of machine and marshalling costs are high however large the installation. It is therefore often concluded that very large machines will be most economical off-shore. Unfortunately, in order to consider the costs of very large machines, it has been usual to scale up present designs and, when these are found to be uneconomic, to discard the idea of very large machines altogether. There are in fact sound engineering reasons why optimum designs are likely to differ for various sizes. For example, three-blade designs are usual at present for medium sized wind turbines and two-blade ones for large, megawatt sizes for which blades represent a larger part of the total cost.

For very large machines, perhaps 200 m in diameter and generating 20 MW or more, completely new design philosophies may well be appropriate. It has always been claimed for vertical axis designs that they can be scaled more economically to very large sizes because they do not have dominant reversing gravity loads to cope with as do conventional propellor type rotors. Both of the vertical axis designs on the CEGB test site at Carmarthen Bay claim this possibility: the "arrow-head" design built by MacAlpines (Anderson et al, 1987) and the "mushroom" design built by Balfour Beatty (Bullen et al, 1987).

A horizontal axis design with the potential of scaling to very large sizes (200 m or more in diameter) is that proposed by Mensforth (1979). The cantilevered blades of propellor designs are replaced by a strutted and hooped rotor reminiscent of the enormous Ferris Wheels to be found at amusement parks around the world. Such strucures are remarkably strong and it is possible to consider rim-drives avoiding the need for a tower. A serious development programme for off-shore wind power should include the investigation of such very large machines.

8. REFERENCES

Anderson, M B, Groechel, K M and Powles, S J R, 1987. Analysis of data from the 25 m variable geometry vertical axis wind turbine, pp333-339, Wind Energy Conversion, Ed J M Galt (MEP, London: 1987)

Bennett, M, Hamilton, P M and Moore, D J, 1983. Estimation of low-level winds from upper-air data, Proc IEE, Vol130A, p517.

Bullen, P R , Mewburn-Crook, A and Read, S, 1987. The performance of a novel shrouded vertical axis wind turbine, pp215-219 Wind Energy Conversion, Ed J M Galt (MEP, London: 1987)

Burton, A L (Ed),1985. Report on UK off-shore wind energy assessment: Phase 2B, Vol 1 (Executive Summary). Taylor Woodrow, Southall, UK. Feb 85. (From ETSU, Harwell, UK)

Hardell, R, 1984. Wind power production off the Swedish coasts: estimate of generation costs. IEA7-B3.03. AIB, Stockholm, December 1984

Hardell, R, 1986a. Technical specification for OWECS prototype. IEA7-C02. 3K Engineering AB, Stockholm, April 1986

Hardell, R, 1986b. Objectives for an off-shore prototype wind turbine. IEA7-C04. 3K Engineering AB, Stockholm, Apr 1986

Iperen, J van, 1980. Conceptual design of an OWECS power station IEA7-B01 Hydronamic BV, Sliedrecht, NL. Feb 1980

Koten, H van, 1986. Study of OWECS: conceptual design of supporting structures. Fugro BV, Leidschendam, NL. Aug 84

Lindley, D, 1988 Economical wind-farming Chartered Mech Eng Feb 1988, pp17-19.

Maribo Pedersen, B, 1983. Off-shore wind power in Denmark. IEA7-0.05 DEFU Report EEU-83-01E. Lyngby, Denmark, Mar 83

Mensforth, T., 1979. Wind power generation on a large scale, IEE Conference Series Vol171, p268 (IEE,London:1979)

Otway, F O J, 1986. Off-shore wind energy: a comparison of British, Swedish and Danish studies. IEA7-B1.3.21 GDCD Report GDCD/PE-B/SS/97, CEGB, Barnwood, UK. June 1986

Palutikov, P J, Davies, A & Kelly, P M, 1985 . The variability of the wind field over the British Isles - the implications for wind power production. IEA-A13 University of East Anglia, Norwich, UK, April 1985 (From TPRD, CEGB HQ, London, UK)

Swift-Hook, D T, 1987a. Wind power for the UK, ISBN 1870064 01 1 (BWEA, London: 1987)

Swift-Hook, D T, 1988a. Fast moving wind energy, Energy World, No155, Feb 88, p13.

Index